Cindynics, The Science of Danger

We can't solve problems by using the same kind of thinking we used when we created them.

Albert Einstein

Reliability of Multiphysical Systems Set

coordinated by
Abdelkhalak El Hami

Volume 11

Cindynics, The Science of Danger

A Wake-up Call

Guy Planchette

iSTE

WILEY

First published 2021 in Great Britain and the United States by ISTE Ltd and John Wiley & Sons, Inc.

ISTE Ltd
27-37 St George's Road
London SW19 4EU
UK

www.iste.co.uk

John Wiley & Sons, Inc.
111 River Street
Hoboken, NJ 07030
USA

www.wiley.com

Library of Congress Control Number: 2021941729

British Library Cataloguing-in-Publication Data
A CIP record for this book is available from the British Library
ISBN 978-1-78630-728-6

MIX
Paper from responsible sources
FSC® C013604

Contents

Acknowledgments

The work in this book was carried out by the "*Cindyniques pour tous*" ("Cindynics For All") Working and Reflection Group of the *Institut pour la Maîtrise des Risques* (French Institute for Risk Management). The term **Cindynics is used to refer to the science of danger**.

This group was facilitated by

Guy PLANCHETTE

Editors and Proofreaders

Laurence BAILLIF

Claude HANSEN

Manuela TANCOGNE-DEJEAN

Philippe HOUSSET

André LANNOY

Philippe LE POAC

Jean-Pierre PETIT

Christophe POIRIER

Jean-François RAFFOUX

Active members of the working group

Annie BERTRAND

Nadia BONNET-HOUDETON

Alizée NICOLINO

Myriam MERAD

Danièle TRAUMAN

Marie-Hélène VERGOTE

Philippe BOULENGER

Yves CABROLIER

Pascal COHET

Jean-Pierre CRESCIONI

Vincent DAUMAIL

Mohamed HAOUARI

Jean-Paul LANGLOIS

Rémi PAROUTY

Olivier POULAIN

Vincent RICHARD

Thank you to all of these people.

Presentation of the *Institut pour la Maîtrise des Risques* (French Institute for Risk Management)

The *Institut pour la Maîtrise des Risques et la Sûreté de Fonctionnement* **(IMdR-SdF)** was created with the support of 14 major industrial groups[1], succeeding the *Institut pour la Sûreté de Fonctionnement* on July 1, 2002.

In the summer of 2006, the European institute *Institut Européen des Cindyniques* (IEC) joined the IMdR-SdF to form the **IMdR**, *INSTITUT POUR LA MAÎTRISE DES RISQUES – Sûreté de Fonctionnement-Management-Cindyniques*.

As a **key player**, the Institute's vocation is to help companies and public bodies to adopt a preventive approach when studying situations where threats and opportunities are intertwined. To do so, the Institute studies various methods and tools to identify, evaluate, quantify, prioritize, control and manage feared events in order to transform possible damage into benefits. This approach is intended to be systemic and global: systemic, because the nature of the dysfunctions or deficits can be technical, documentary, organizational, behavioral, environmental and so on; global, because it must encompass all organizational functions (technical, financial, legal, etc.).

Bringing together the skills of both associations, the IMdR now offers its members a forum for exchange on risk management, from political and strategic visions to concrete, everyday scientific tools.

---

1 Air Liquide, CEA, CNES, Dassault, EADS, EDF, Ligeron, PSA, RATP, Renault, SNCF, Technicatome, Thales and Schneider.

The qualities of the IMdR are based on:

– the dynamism of the 20 or so of its working groups;

– the elaboration of studies carried out in the form of projects commissioned by industrialists;

– meetings: theme days or breakfasts, the congress on risk management and operational safety (Lambdamu) and, alternately, the congress with a cindynics theme (*Les Entretiens du Risque*);

– an Orientation, Documentation and Information Center (*Centre d'Orientation, de Documentation et d'Informations*, CODIT) that welcomes and guides through all of the specialized documentation;

– the quality of its relations with all of its partners.

The advantages of the IMdR include:

– its capacity for sharing acquired over more than 20 years, thanks to its experience in bringing together several companies around studies chosen by them. Thus, the Institute offers them the guarantee of knowledge sharing, lower financial investment and partnership;

– its monitoring and benchmarking approach, which makes it possible to have knowledge of the methods used by other sectors, as well as the analyses established following serious events;

– its network of experts, which is based on their high existing potential in terms of dependability and cindynics, both in the industrial and service sectors and in university laboratories. These experts bring a real dynamism to the IMdR's 20 or so working groups;

– its symposia, study days and training courses, which allow a wide dissemination of ideas and concepts of risk management in a warm atmosphere facilitating the exchange and appropriation of knowledge;

– its technical information center, which holds a large library capitalizing on years of studies on dependability, risk analysis, risk control and cindynics;

– its openness on a European level, thanks to contacts with European associations dealing with the same subjects.

Foreword

For many years now, the latest books on cindynics have been published by the publisher Economica: the French books "*Cindyniques – Concepts et mode d'emploi*" [KER 07], and an application to the health sector "*Cindyniques et santé – Contributions des sciences du danger à la santé*" [FES 09]. In other words, this book is welcome because, since the 2000s, cindynics have been the subject of numerous studies and research activities, both at the *Institut pour la Maîtrise des Risques* in the framework of the working and reflection groups "*Les cindyniques à la portée de tous*" and "*Gestion des risques, cindyniques et nouvelles approches systémiques dans le secteur de la santé*", and in the academic world, at the Université Paris 5 and the École des Mines de Paris, to mention a few.

This book was written by experts from different sectors of industry, health and transport. It was an ambitious goal to achieve it and it has been achieved. We would like to thank them for this work and for its publication.

The text includes a very useful glossary explaining the difficult words or expressions used in the vocabulary of cindynics, such as deficit, dissonance and activity situation, among many others.

Under the heading hazard, the book points out that the ISO 31000: 2018 standard uses the term risk source to qualify the hazard. In practice, the terms hazard and risk are often used interchangeably. However, they are two quite distinct terms. According to the Longman Dictionary of Contemporary English (1978, p. 278), *hazard is the possibility of harm or loss*, which can threaten safety/security. Hazard is a potential, it expresses a possibility, which may occur depending on conditions, depending on the context, which may not necessarily be realized, something virtual. Risk is the likelihood that exposure to a hazard will lead to a negative consequence. It is measured by the coupling of the values of the likelihood of a given danger and the seriousness of the consequences of this hazard. Quite simply, there is no risk if there is no exposure to a hazard. Risk therefore has a negative image. However, the standard has introduced the positive consequence aspects of risk. "To win you have to risk losing" (Jean-Claude Killy).

The foreword is based on a curve of accident rate trends from 1987 to 2016, showing a decrease in accident rates until 2003–2004, followed by a seemingly asymptotic constant plateau, a tango on the asymptote [FRA 04]. Technical advances have allowed this rate to decrease over the years. Today's accidents show that we have now reached a threshold that would be insurmountable, calling into question the organization of work. We can now see that all too often the failure to take into account feedback and the lack of understanding of physical phenomena are two deficits frequently revealed by analyses of recent technological accidents.

The organization is a complex system and all vulnerabilities must be identified.

The cindynics approach seems well armed to deal with this problem.

Chapter 1 is devoted to the understanding of cindynics. It defines the notion of activity situation by its contexts and by the activity (a movement), characterizes it, detects evolutions, whether observed or not, and evaluates gaps or inconsistencies. In the end, we are very close to the practice of a HAZOP method or imaginative, but knowledge-based investigation, akin to Sherlock Holmes.

The hazard hyperspace represents the five characteristics of risk sources: data, models, regulations, values and purposes. The process consists of conducting interviews with experts, consulting existing documentation or feedback, analyzing the content and structuring potential risk sources, identifying and analyzing gaps, and establishing a synthesis matrix known as the deficit and dissonance matrix. This is probably the strong feature of cindynics. Thanks to this analysis based on the five characteristics above, it is possible to highlight all of the sources of risk, particularly those related to the organization, which is not necessarily possible in the usual analyses.

The applications at the end of the book show that the approach is fully applicable for *a posteriori* analyses. The authors believe that it is also applicable *a priori*. However, this may be more difficult. For example, for the analysis of an innovative product, regulations do not exist, few data are usable and models may be non-validated or non-existent.

Chapter 2 describes the methodology:

– first, activity situations are characterized. It is a question of specifying the temporal horizon, the spatial and geographical horizons, the context and the networks of actors;

– dangerous situations are constructed and then qualified (description, highlighting of dysfunctions, pathogenic elements, deficits and dissonances).

The authors believe that it is possible to limit situations in space-time. Therefore, this approach has an advantage over classical risk analysis, which is only a snapshot at a given moment; we are therefore obliged to redo the risk analysis periodically in view of changing conditions, context and functioning.

Chapter 3, particularly well developed, presents the hazard hyperspace grid, which is the qualification tool. It contains the five characteristics mentioned above. The analysis of major accidents shows that there are several causal chains called SCDs or systemic cindynogenic deficits: four cultural (conviction of absence of peril, unfavorable to complexity, no-communication attitude, no-attention to the outside world), two organizational (production dominates risk management, dilution of responsibilities) and four managerial (absence of a feedback policy, no-risk management methodology, no training for risk management and safety personal, insufficient knowledge and lack of preparation for crisis management).

Chapter 4 provides decision support. It builds the deficit matrix and the dissonance matrix from which a decision-maker will decide whether or not to mitigate hazards, with the aim of minimizing the consequences of deficits and dissonances. We are not sure that this is an optimization. The decision-maker lacks quantitative benchmarks on both the likelihood and magnitude of the consequences. Their decision could lead to cost overruns or security under the target. This seems to be the weak point of the approach, which can certainly be improved.

In conclusion, the authors believe that the cindynics approach is mature and allows us to understand and explain how dangerous situations are created. We also think this for *a posteriori* analyses in view of the three applications discussed in the book. The authors state that *a priori* analysis is possible, which is highly probable, but we think that the aid to the decision-maker should be reinforced. This can only be done if we move towards quantification.

At the end of the book, three technological accidents are dealt with: the Bhopal disaster, the Queen Mary II footbridge and Deep Water Horizon. The reader will find the direct application of the chapters of the book. The results highlight the deficiencies, the main risk factors, for example, for Deep Water Horizon: production pressure and lack of physical understanding of the phenomena; for Bhopal: conviction of absence of peril, lack of feedback and so on, but no mention is made of toxic gas dispersion models.

Several perspectives are proposed by the book. It seems to us that the priorities are the following:

– to develop applications, in any sector;

– to develop links with the SWOT method (Strengths-Weaknesses-Opportunities-Threats, which is a business strategy tool for determining options and identifying factors that are favorable or unfavorable to the achievement of objectives);

– to help the decision-maker choose options to reduce risk;

– to develop and analyze complexity in the design phase.

The cindynics approach is clearly applicable, it is mature, and it can be used to better understand the organization's risk factors, to better approach a design process.

This book is, at the same time, a state-of-the-art of knowledge, a synthesis and a working tool. It should be of interest to a wide audience of actors involved in risk management. It is a good answer to many topics for managers, project managers, design engineers, feedback analysts, authorities, researchers and PhD students, among others, in short, for all practitioners, whether they are beginners or experienced.

André LANNOY
Vice President of the *Institut pour la Maîtrise des Risques*
July 2021

Preface

Danger existed long before man, who, in order to survive, was forced to face it by taking risks.

A bit of history

Ever since man has existed, he has always been confronted with dangers that he had to face if he wanted to survive, leaving him mostly helpless and afraid of the forces of nature. Ancient populations, considering dangers as a divine punishment, devoted themselves to oracles [DES 03].

Later, the needs of development, domination and trade brought new forms of danger. Towards the end of the Middle Ages, the notion of risk emerged when, in order to obtain a better situation, humans sought to confront danger. It was then considered as an accident that man could try to prevent. It was only on the basis of the geometry of chance, initiated in 1650 by Blaise Pascal, that a first stage of probabilistic studies led to the establishment of mortality statistics [LEB 00]. Since the 18th century, theology and philosophy could not justify such a manifestation of divine anger, it was only after the earthquake that struck Lisbon in 1755 that the notion of **responsibility** for the positive and negative consequences of personal decisions appeared. The concept of **risk** was then born, and gradually, damage ceased to be perceived as divine punishment.

To counter the danger, each human being, each community, and then each profession has therefore invented, adapted and perfected solutions according to its specificity, the history of the nature of the difficulties encountered and the types of risks.

However, since its appearance at the end of the Middle Ages, the **notion of risk has gradually obscured the concept of danger** because it "has been progressively enriched by its contact with the disciplines that it has invaded, namely in turn probabilistic calculation, game theories, economics, engineering and technological risk control, psychology and risk perception, and sociology" [KER 11].

All of these developments in the area of risk have led to a multitude of concepts and methods[1] that make it possible to achieve safety adapted to the context of activities, better operating safety of the equipment used and an ability to assess and control risks.

However, have we really lost awareness of the notion of danger?

For example, let us look at recent milestones of progress:

1950 to 1970

The increasing number of technical failures in electronic equipment prompted a great deal of research.

Although they existed long before our era [LAN 18], approaches to operational safety and risk control were structured from the 1950s onwards. New technologies (high-performance weapon systems, telecommunications, etc.) were developed, in particular with numerous forms of electronic equipment. The reliability of new equipment could not be compared to that of mechanical equipment, leading to an increase in technical failures. Based on this observation, many research programs were launched within companies operating in the fields of telecommunications, nuclear and space. After these years of research, the overall number of accidents was considerably reduced between 1950 and 1970.

1970 to 1986

In the face of a series of disasters, research focused on human failures.

This period was unfortunately marked by a series of technological disasters (Flixborough, Seveso, Three Mile Island, Bhopal, Chernobyl, Challenger). These unexpected crises highlighted the fact that human-generated failures at the operator level had to be added to technical failures. This new consideration also contributed to the reduction in the number of accidents and disasters.

---

1 Consult the IMdR website: https://www.imdr.eu/offres/doc_inline_src/818/Fiches_methodes_m2os.pdf.

1986 to 1990

Research continued around the notion of a system and the living organism model. One observation emerged by analogy with the weakening of the immune system. Cindynics was then born.

Evaluating the importance of this cascade of events, Lagadec published a book in 1987 entitled "*Le risque technologique majeur*" [LAG 83][2]. At the same time, Kervern, with the help of a team of researchers, set out to understand the reasons for this chain of failures. They studied a series of accidents, whether technological, natural or domestic, and deduced that **the causes were not only due to technical failures, but also to those of the human operator.** These feedback studies revealed strange similarities with the notion of immune "deficiency" in living organisms. By analogy, they concluded that an accident occurs when pathogenic elements (such as gaps, vagueness or ambiguities) are created within organizations and do not encounter "appropriate defense systems". **This discovery reinforced the primacy of the concept of danger over that of risk**, and the name **deficit** was given to these pathogenic elements. This is how the science of danger or **cindynics** (name taken from the Greek *kindunos*) was born, bringing to light, at system level, the notion of **systemic cindynogenic deficit**. And, these pioneers, from all of the "small islands" explored, were thus able to identify the **archipelago of danger** [KER 91]. The resulting book, "*L'archipel du danger*", has a postface by Laborit, a medical surgeon and neurobiologist, who was one of the pioneers of complexity theory and the initiator of complex thinking.

The small islands of the archipelago (probability calculations, data processing, actuarial calculations, financial mathematics, psychology, sociology, phenomenology, ontology, etc.) were already sufficient in number to justify the plural displayed in the term cindynics.

Here, the necessary awareness of the danger was regained.

After 1990

New types of accidents and disasters were calling into question the organization of work.

This period saw new disasters causing real damage to the environment (pesticide factory 91, Mede 92, oil tanker wrecks 92-93-96-99). These events reinforced the

2 See: http://www.patricklagadec.net.

need to complete the list of technical and human factors with that of organizational factors.

However, despite all of these advances leading to spectacular progress in risk management, the downward curve in the number of accidents seemed to have reached an asymptotic limit which, for the time, remained a brick wall. Figure P.1. illustrates this asymptotic limit[3].

One reason may explain this limitation.

A close look at the characteristics of current risk management studies shows that they are based on five pillars: the primacy of analytics, the notion of risk, the dreaded random event, quantitative aspects and finally, those related to the senses. These five fundamentals resemble axioms that will generate certain biases (see Appendix 1).

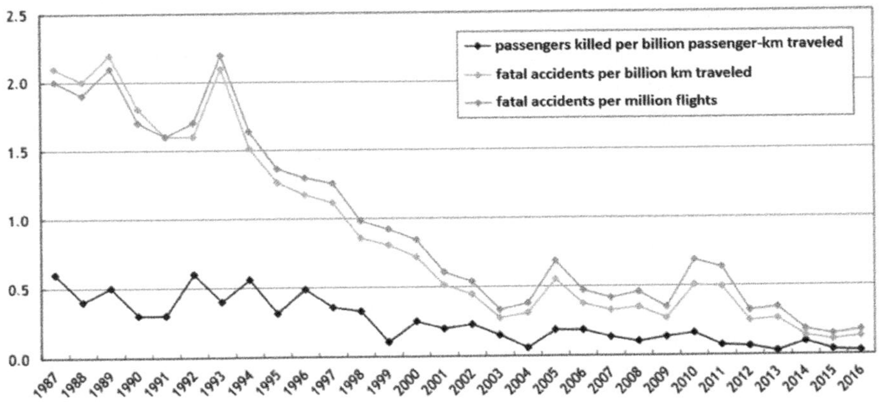

Figure P.1. *Changes in annual passenger fatality and death rates in scheduled services since 1987, Source: Ministère de la Transition Écologique et Solidaire (Aviation Safety Report, 2016). For a color version of this figure, see www.iste.co.uk/planchette/cindynics.zip*

Consequently, based on these characteristics, the description of the potentiality of the events generating risk is based only on a space limited to the two dimensions noted (p,g), characterizing the probability of occurrence p of a dreaded event and the severity g of the consequences generated.

3 Aviation security report, 2016: https://www.ecologique-solidaire.gouv.fr/sites/.../rapport_securite_aerienne_2016.pdf.

Not giving up on pushing back this limiting wall, Kervern and his team found that human beings are at the heart of organisms and fulfill diverse and varied roles. Their goals are both complementary and antagonistic. Moreover, the existence of these actors intervening within different organizations (see Glossary) generates interactions (see Appendix 2) through the exchange of data, information and so on, leading to the emergence of verbal and non-verbal influence relationships (conflicts of interest, rivalries, cooperation and competition).

Indeed, according to its sociological definition, an organization generates a double problem: it is at the same time a place of creation of social links and a place of actions turned towards the achievement of concrete results. This bivalence can generate tensions, considering for example:

– dissatisfaction, to a greater or lesser extent, with the choices made in terms of the distribution of activities;

– individual recognition needs.

An organization is therefore a place where both ambiguity (source of uncertainty) and the search for a minimum of coherence necessary to achieve the desired goals coexist.

In the book "*Comportements humains et management*" [ALE 09], Alexandre-Bailly *et al.* suggest that organizations are complex systems structured according to three parts: the organizational devices, the culture and the games of the individuals composing the organization.

Therefore, we are faced with a second reason for the limit mentioned above. Because of the presence of these interactions, we do not live in a world where cause and effect relationships are linear. **Indeed, the integration of human and social interactions leads us into the universe of complexity, which can in no way be translated by a two-dimensional space.**

As a result, any actor or group of actors (see Glossary) with their own characteristics may cause other types of pathogenic elements, such as ambiguities, blurred, divergent points of view, conflicts and rivalries.

Thus, in addition to deficits acting as gaps, dissonance is also a source of tension between actors. Other forms of danger (and therefore dangerous situations) than those identified up until now also appear. It is therefore necessary to better define the danger, that is, "to identify it, in order to be able to hope to reduce its negative consequences" [KER 91].

Faced with this awareness, it became crucial for Kervern to find another mode of description capable of identifying the potentiality of these new types of dangers, a mode that could no longer only be expressed by the dimensions mentioned above.

Fortunately, even though dangerous situations defy human perception, humans have a certain capacity to undertake the description. It was the art of establishing a cindynics epistemology which, based on the work of Simon, Le Moigne, Morin and Mugur Schächter, made it possible to proceed, by describing dangerous situations through a process of genesis of danger (see Chapter 1, section 1.2 and Chapter 2).

These sciences of danger thus widen their fields of investigation by seeking in priority, all of the deficits and dissonances that can generate vulnerability in the organization. Because of this vulnerability and under the effect of pathogenic elements, the dangerous situation thus created tends to evolve, either progressively or abruptly towards accident, catastrophe or crisis.

These cindynics concepts have been transferred from the industrial sphere to the civil society sphere, with applications to the family [FOU 97], the city and public health. After September 2001, these concepts explored the transition from industrial and psychological risk to threat. And other perspectives are being developed to deal with other aspects of conflict (see Chapter 6).

These developments show the relevance of the use of the risk analysis approach and its adaptability to any complex situation dealing with hazards and risks.

The birth of a new science: cindynics

In October 1986, the specialized journal "*Les Annales des Mines*"[4] devoted its entire issue to major technological risks. Five major chapters covered: risk assessment methods, the industry faced with major risks, prevention and regulation, preparation and management of crisis situations, society and major risk.

At the beginning of 1987, the French Association of Executives for Social and Economic Progress (*Association Française de Cadre Dirigeants pour le Progrès Social et Économique*, ACADI) chaired by Kervern, organized a series of meetings bringing together experts from different backgrounds: chemistry, nuclear, aeronautics, oil and so on. From these meetings, the idea was born to organize an international conference on the subject. ACADI was joined by the magazine "*Industries et Techniques*" and the insurer UAP. An organizing committee, bringing

4 Founded in 1794, this scientific journal with an international audience is published in English, German, Spanish, Russian and French.

together the best French experts in risk management, worked hard for several months to explore this new scientific field. It was led by Michel Horps, one of the directors of Crédit Agricole, and Francine Londez, a specialist in the design of large technological meetings.

Thus, on December 7 and 8, 1987, 1475 people from 13 countries, representing 30 industrial sectors, 320 companies and 90 universities or research centers, met at UNESCO in Paris to compare their experiences and try to define a common policy.

About 100 high-level conference-debates show the real interest of the participants in what was beginning to be called "the science of danger".

Seven commissions met to address:

– progress in safety in large, complex systems (lessons learned from major industrial accidents, risk control in large companies, reliability/organization/systems-expertise/decision-making);

– the human factor and safety (typology of human failures, tools and methods to integrate the human factor);

– management of global environmental risks;

– the preparation and coordination of the treatment of technological disasters;

– progress in safety in diffuse systems (domestic and transportation risks, workplace accidents);

– the economic and financial aspects of safety (prevention, insurance, European standardization);

– the collective perception of risks and accidents.

At the closing session, a certain number of commitments were made. Among them, to give a name to this new discipline and to promote its teaching throughout Europe. An institute (*Institut Européen des Cindyniques*) created for this purpose was set up in Paris in January 1990.

The birth and activities of the IMdR working group

Kervern's powerful and modern intuitions have allowed an understanding of the genesis of pathogens in numerous situations where many actors exist. However, his writings, changing thinking habits, were not easily assimilated, especially since the vocabulary that was used was unusual.

The *Institut pour la Maîtrise des Risques* (IMdR) therefore considered it essential to create a working and reflection group to **make cindynics available to everyone in order to share their interest and usefulness, and to show that this approach is particularly well adapted to the complexity of today's world**.

Working group members with different professional backgrounds (engineers, safety professionals, teachers, researchers, etc.) pooled their experiences to achieve these objectives by meeting between 2013 and 2017 on five afternoons per year. The general idea of the participants was to put themselves in the position of someone who wanted to use the cindynics concepts without knowing how to do it.

Initially, the approach consisted of deciphering, simplifying and sharing the terminology used to define cindynics concepts, without distorting their spirit.

In a second step, the members of the working group applied risk analysis to an example of a disaster (Bhopal, see Chapter 8) and proceeded with an in-depth analysis of the approach.

This study first learned the need to have **sufficient information** to establish the context of an activity situation (see Chapter 2, section 2.2) to be analyzed, to identify the horizons that will characterize the scenario in which the drama unfolded, and to make a judicious choice about the actors shaping the situation.

During the various information collections, the learning process consisted of **carrying out successive analyses of the content of the data collected** (see Chapter 2, section 2.3.2). This work of qualitative data[5] analysis used a research method from the social sciences "aiming to build theories, not from predetermined hypotheses, but from collected data" (see Appendix 3). This method allowed all of the elements of the activity situation to be progressively built in a more exhaustive way.

In addition, it has made it possible to specify the order of the stages of the cindynics process. The results obtained led to the understanding that before proceeding to the causal study of the event initiating the disaster, it is preferable to start from the study of a dangerous situation creating vulnerabilities, thus offering its fragilities to any disruptive event.

This exercise also highlighted the difficulty of being confronted with the complexity of the universe being studied and of going beyond the traditional framework of "cause tree" studies, starting from the event at the origin of the

5 Grounded theory (Glaser and Strauss, 1967), adapted by [PAI 94]. See: https://doi.org/10.7202/1002253ar.

accident. This discovery was expressed by one of the participants: "in short, the cindynics approach appears to induce a particular posture of attention to the elements of the situation, beyond the physical system". They became aware that the **context in which the event occurs is as important to study as the event itself.**

Faced with complexity where everything is the cause and vice versa, the group was confronted with a change in thinking: because this complexity requires a looped thinking process, taking into account the different levels of organization.

At the mid-point of the group's work, it was deemed useful to reflect on what the participants had learned and retained during the working group meetings:

Ten responses were received to a questionnaire addressed to each of them. They were first processed using the content analysis method (see Appendix 3). This made it possible to draw up the following synthesis. This synthesis highlighted the differences in approaches and needs according to the length of participation in the working group and prior knowledge of cindynics concepts.

What was seen as important and rewarding:

– **Concerning the work of the group**:

- For five of us, the plurality of the points of view expressed appeared pleasant.

- For four, it was the atmosphere: good mood, sustained exchanges (even if they were sometimes too long), freedom of speech.

- For three, the methods used seemed effective: the principle of work in sub-groups, the use of tables and hyperspaces of danger (see Chapter 1, section 1.3), the definition of horizons and actors, regular reports, theoretical contributions, periodicity… and especially the work on real cases.

- This approach stimulated personal work outside of the group, either as an extension of the group, by participating in other places of reflection, or by taking the theory further by consulting key works.

- On the contrary, there was the impression that the group was venturing along several paths without really concretizing any of them, because the initial objective was to prepare a document to popularize cindynics, which seemed to have been abandoned.

– **Concerning the work on cindynics**:

- By focusing on the human and the organizational elements, the difficulties of interpreting the information to be classified on the five axes of the hyperspace of danger were great (4). They posed the problem of expertise that would be necessary.

- The cindynics approach was reinforced (5), because it allowed in-depth work (3), widened the perception of risk by taking into account the notion of danger (3), emphasized the importance of the context – horizon or situation – (2), but above all by the fact that it proceeded by successive adjustments (2) by emphasizing the human and the organizational elements, which refined the understanding of past phenomena that have led to the sequence of causes of an accident.

- Cindynics concepts thus made it possible to go beyond the only questioning on the reliability of the equipment.

- Moreover, the analysis, established from the danger hyperspace, had the advantage of an organization testing the coherence existing between the five aspects of the danger hyperspace.

- However, it involved difficulties: when studying complex systems (3), within which the actors themselves had complex relationships – implicit or unknown – (2), it required working on precise data with a clear delimitation of the scope of the study (2) and the actors to be taken into account (2). However, there were few real cases studied that could serve as reference (2).

– **What it seemed essential to take further**:

- The question of the production of work "outside" of the group was again raised (5), either in the form of a glossary, or by resuming the work done on Bhopal, or in the form of teaching students (initial or continuing education).

- Conversely, the other question was working on a real case proposed to the group by an external authority (2), in order to question the predictive power of a cindynics study.

– **In the search for deeper knowledge**:

- A more precise study on deficits was requested.

- A critical reading by the group of an article already written by one of the members.

- The search for a computer tool to facilitate the task of data exploration and exploitation.

Subsequently, members of the group were interested in studying a case presenting dysfunctions, but not yet having caused an accident, with a view to using the approach in the field of prevention. However, this work could not be completed in the time allotted. Indeed, the insufficiency of the available information made it impossible to correctly establish the context of the activity situation and to thus study the deficits and dissonances that could exist within this chosen situation compared to the situation deemed satisfactory by the study's sponsors. On the contrary, this example highlighted the real difficulty in choosing the "networks of actors" working within the situation or in integrating a team of cindynicians into a project from the design phase. It also showed the need to conduct interviews with the actors deemed relevant, in order to record their knowledge and perceptions of the situation.

Subsequently, other examples of accidents were studied and the results of the different experiments and work were compiled in a collective file (see Chapter 8). Even though the studies were sometimes incomplete, they nevertheless prove that cindynics concepts can provide a better response to our problems of dangers, threats, conflicts and risks. Hence, the members of the working group wished to disseminate their work in order to demystify cindynics concepts, which are neither an obscure science nor reserved for a few specialists, and to show that, on the contrary, cindynics are adapted to the complexity of today's sociotechnical or societal systems.

In view of the difficulties encountered, it would seem that, more than the obstacles related to vocabulary, it is indeed the amplitude of this new discipline of thought that may have put off the very first aspirants to cindynics concepts, who tried to embark on the adventure without sufficient preparation or clarification of the process.

July 2021

1

Understanding Cindynics

While the methods, studies or standardization commonly used to manage risks have been quite effective, it is undeniable that through the history of accidents and disasters, their effectiveness is still not entirely satisfactory.

In order to fully understand cindynics, it seems useful to examine what its concepts can add to postponing the occurrence of feared events.

Indeed, any risk study currently conducted starts with the knowledge of the dangers existing within the situation under examination. This phase, which is the keystone of a risk study approach, is called the "hazard study".

Let us recall that the method most often used for risk studies is the one called MADS–MOSAR which involves:

– a model called "Methodology of Systems Dysfunctioning (MADS)" also called Universe of Danger [PÉR 00];

– a method called MOSAR, "organized method of systemic risk analysis" that allows the identification of major risk scenarios using methods and tools for operational safety.

The use of the MADS model makes clear the priority given to the search for hazards[1]: it uses "systemic modeling by breaking down the installation into sub-systems and systematically searching for the hazards presented by each of them".

This model is used to establish a hazard qualification grid.

1 The MADS–MOSAR method (Methodology for System Dysfunction Analysis – Organized Method of Systemic Risk Analysis) is a risk analysis method developed by [PÉR 00]. By referring to the words "system" and "subsystem", the approach is deemed systemic.

Concepts linked to cindynics are therefore perfectly aligned with the **principle of thinking about danger before paying attention to the risks and the initiating event**. However, the hazard research carried out by MADS is limited to the exploration of a table of hazards of technical origin[2]. And while these have been brought to our attention, it is because they have been described and catalogued, because they are perceptible. It is therefore on this level that cindynics concepts differ and bring about very great originality, because the dangers or sources of risks are certainly of technical origin, but they are also organizational, cultural or managerial (see section 1.1).

Cindynics therefore meets the need for in-depth research into complementary sources of risk, either difficult to identify or coming from pathogenic elements which, if they do not attract attention, can unknowingly weaken an organization's ability to resist hazards (see the term "organization" in the Glossary)[3]. Recalling the importance of focusing on sources of risk that are difficult to perceive, [DEL 20] notes: **"the two-week fall in the Dow Jones and the CAC 40 gives credence to the famous black swan theory, developed by the statistician Nassim Nicholas Taleb, according to which the greatest financial catastrophes are triggered by seemingly minor events whose rarity is such that it makes them unpredictable for market operators".**

The ISO 31000:2018 standard states that the success of risk management depends on the effectiveness of an organization's management framework integrating the understanding of the organization and its context, the establishment of risk management policy, responsibilities and so on, in order to lead to a risk management process.

It further states that:

– managing risk is part of governance and leadership and how the organization is managed;

– managing risk includes interaction with stakeholders as an integral part of all activities of the organization;

– managing risk considers the internal and external context of the organization, including human behavior and cultural factors.

However, while the cindynics approach is completely in line with the prescriptions of the ISO 31000:2018 standard, it obligatorily precedes the risk

2 See the box on page 8 of [PÉR 00].

3 The examples of accident and disaster studies, detailed in Chapter 8, illustrate the capabilities of cindynics to explore these complementary risk factors.

assessment phase by the study of an assessment phase of all types of hazards (or sources of risk).

A "risk mapping" tool is commonly used in risk management in order to create and share a risk frame of reference.

Cindynics tools also use a frame of reference which concerns sharing through a process that is in line with the recommendations of the ISO 31000:2018 standard but whose purpose is to highlight the **hazards** through matrices of deficits and dissonances.

NOTE.– In the remainder of this book and to comply with the ISO 31000:2018 standard, **the term "risk source" will be used instead of "hazard".**

Then, to become fully aware of the originality of cindynics involves discovering that its concepts incorporate a method, tools and processes that are adapted to the complexity of systems, as well as an approach to the study of risk sources adapted to the complexity of systems.

1.1. The approach

During its life, any organization receives many internal or external requests (desired development, opportunities, imposed constraints, changes in objectives, realization of new projects, reactivation of a process, changes in regulatory markets, staff turnover, sociological changes, etc). As a result, any organization modifies itself voluntarily or not with time **through these solicitations called transformation operators**.

These inescapable developments, whether desired or endured, generate as many potential **weaknesses as strengths (physical, organizational and psychological)**.

The cindynics approach therefore primarily monitors the fragility of collective activity situations, that is, the emergence of dangerous zones (see section 2.3.1) within an organization, thus weakening its capacity for resilience. These fragilities, due to the transformation operators, lead to deviations from a nominal or desired organization (state of the art, regulations, standards, charters, procedures, data and culture). The tensions generated by these gaps become risk sources that can lead to incidents, accidents or disasters.

This approach is therefore similar to that of physiopathology, a discipline that deals with disturbances in the normal mode of functioning of the constituent elements of a living organism.

All of the originality of cindynics lies in this ability to detect these sources of particular risks produced by these imbalances. And if, in doing so, potential forces are detected, it would be a pity not to take advantage of them.

When it comes to analyzing technical risks (which are **describable**), the methods and tools of dependability are perfectly suitable.

On the contrary, other categories of risk sources, more related to human, organizational, managerial, cultural and environmental aspects, being more difficult to identify, receive much less attention in order to be identified and described. As a result, cindynics, by placing the human and the organization at the heart of the complexity of organizations, focuses first on these categories of risk sources and then uses the MADS-MOSAR method to deal with technical risks.

Thus, the notion of an **activity situation** has become the **founding concept of cindynics** because it obliges us to define what is necessary to talk about when approaching a hazard study. The terms "situation" and "activity" still need to be defined:

– "situation" is to be understood etymologically as "the location of a structure". It is also, according to the Larousse definition, the set of events, circumstances and concrete relations in the midst of which someone or a group finds itself;

– the word "activity" is borrowed from the Latin "*activitas*" or "*vis agendis*", "the will to do". Later, the meaning of "activity" became "exercise of the power to act", then movement.

This is, indeed, the meaning given to the **activity situation which integrates notions of history, geographical space, actors and movement, and therefore dynamism and development**.

1.2. The method

Based on the fundamental concept, namely the activity situation, the resulting cindynics method is based on two notions as follows:

– **the characterization of this activity situation** by an objective approach of the internal and external contexts[4], complemented by other observable criteria (see section 1.3);

– **the qualification of hazardous areas** within it, whether directly observable or not. It is therefore imperative to **acquire the ability to describe** unobservable

4 See ISO 31000:2018.

elements. This ability was brought about by the scientific advances of the beginning of the 20th century. Indeed, since then, the reality of the world has evolved considerably thanks to two fundamental discoveries: **quantum theory**, which has brought to light the existence of invisible aspects of matter, and **complexity**[5].

Consequently, in order to succeed in describing the unobservable elements, cindynics adopts **a method of description** known as the "method of relativized conceptualization (MRC)" [MUG 77] proposed by Mioara Mugur Schätcher, a specialist in quantum physics. This method, born from the inability of physicists to observe – by means of measuring instruments, even very sophisticated ones – the invisible phenomena existing within matter, is based on a new process of constructing knowledge of the unobservable.

Through repeated observations, descriptions of the **impalpable or unobservable** are obtained by producing more or less objective images called **"aspects"** to ensure qualification.

The multiplication of qualifications carried out by observations in arbitrarily large quantities leads to a qualification grid used to establish a **"relativized description"** of the unobservable.

This method of describing risk sources that are not easily observable thus makes it possible to identify and evaluate the gaps or inconsistencies generated by the inevitable changes that organizations experience[6].

We have already noted that the original organization cannot be considered intangible. It must be adapted to this inescapability in order to prevent any possible dysfunction. This notion of **temporality** is all the more important since it is often slow, sometimes almost impalpable, evolutions that are at the origin of disasters. It is therefore a question of **integrating the notion of transformation operators** (see Chapter 2, section 2.1).

1.3. The tools

The set of tools used by cindynics identifies the devices used to characterize activity situations and to qualify areas considered dangerous:

– the characterization of situations is performed through criteria relating to five domains (spatial, geographical, chronological, contextual and networks of actors) (see Chapter 2, section 2.2);

5 The development of quantum theory will be briefly discussed in Appendix 4.

6 The use of the description method will be discussed at length in Chapter 8.

– the qualification of hazardous areas cannot be limited to the technical qualifier alone, it broadens its scope by selecting five others. For, in order to take the other risk sources into consideration, it appeared indispensable to report on the aims of the collective work, the methodological aspects, and the philosophical, ethical and deontological questions, specific to any organization and, by definition, complex.

Hence, in addition to the scientific "aspects" represented by **data** and **models**, cindynics has added complementary qualifiers relating to the humanities, integrating the **regulations** or **rules** undergone or imposed, the **values** that guide actions towards goals and **purposes** that are the aims of the organization. Thanks to these five **qualifying aspects**, risk sources that have not been identified so far become accessible to the description[7].

Laws are acts passed by Parliament. **Regulations** emanate from the Executive and are expressions of laws. Their applications are imposed. By extension, within an organization, rules are the expression of a decision from different levels of management. It is understandable that these notions also apply to the term procedure.

In contrast, standards are voluntary. Compliance is not mandatory.

With respect to **values**, this term refers to attributes and perceptions that a person shares with members of his or her social or cultural group. Types of sociological values include moral and ethical, ideological (political), spiritual (religion), beliefs, for example, ecological beliefs, or aesthetic, open, individual and collective beliefs.

As for the term **purpose**, it represents a set of statements, stemming from one or more ideas, which will come in the form of a goal to be reached. For an organization, the purpose is its raison d'être.

It is important not to confuse the term purpose (or goal) with the term objective, because the latter is a step, a means to an end, to arrive at the purpose.

In cindynics literature, this representation of qualification is called the hyperspace of danger [KER 95].

1.4. Processes

Cindynics processes (see Chapters 2 and 3) integrate for an activity situation:

7 The accessibility of the description is presented in Chapter 8.

– its characterization in order to fully understand all the dimensions of the context in which the history of the situation developed and the way in which the behaviors of the various stakeholders developed;

– the qualification of hazardous areas within it; to qualify hazardous areas (see Chapter 2, section 2.3.2) is to practice as many observations as necessary on the five qualifiers of the "hyperspace of danger". These observations are obtained through **interviews and desk research**, selected as the study progresses;

– interviews are conducted with stakeholders[8] identified as relevant to their role or function at the time of the study and/or before (whether internal or external to the observed situation); the documentary references cover all the horizons selected;

– the structured description of the sources of potential risks; this is carried out through an **analysis of the content** (see Chapter 2, section 2.3.2) of the documents collected (interviews, documentary research and observations). This analysis enables the information to be grouped together by dividing it, according to its nature, by type of qualifier so as to **reveal gaps between an organization taken as a reference and the one observed**;

– the identification, resulting from the analysis of these gaps, of the various deficits, whether systemic or individual, and of the dissonances, which are sources of tension between actors;

– the establishment of matrices of these deficits and dissonances; these matrices (equivalent to the design structure matrix (DSM)), for use by the sponsors, summarize these gaps (deficits and dissonances);

– the study of the relevance of the gaps observed; this is carried out with all the partners concerned in order to determine the extent to which these gaps constitute potential vulnerabilities or strengths.

In conclusion, while cindynics was conceived through feedback from numerous accidents and disasters, it has currently succeeded in structuring a scientific approach *a priori* of risk sources.

It now has the ability to **diagnose**, understand and **assess the potential vulnerabilities of an organization before an accident**. This diagnosis, informed by studies of relevance of the potentialities, then makes it possible to **imagine and define with the decision-makers measures for prevention if the risk is future-based and for treatment if the risk has already manifested itself.** These measures will consist of a **plan to reduce** or even eliminate the fragility of the organization.

This plan must **aim both to adapt the practices of the actors concerned and to adapt the organization that served as a hypothesis at the start of the study.**

---

8 As defined in the Glossary.

2

The Usefulness of the Cindynics Approach and Method

Let us recall here the capital importance of exploring, prior to any investigation on risk, the state of vulnerability of activity situations.

2.1. The situation, the founding concept of cindynics

Any situation evolves inexorably with time through transformation operators (see Chapter 1, section 1.1). Kervern and his team then relied on the **concept of propensity,** "an inherent tendency to change that manifests itself in all of the entities of the dynamic universe in which we live" [BEL 95, p. 203]. They concluded that the **inevitable evolution of situations generates a propensity to direct them towards the emergence of a favorable or unfavorable outcome.**

Therefore, the **conditions conducive to this transformation are potentially in place in the activity situation.** Consequently, for the unfavorable aspects, it is not just a matter of avoiding the occurrence of a feared event (preventive action) or reducing its consequences (corrective action), but of **preventing the activity situation from diverting under the effect of risk sources that occur without our realization.** The detection, identification and characterization of the transformation operators find their importance in an anticipated and proactive approach to the analysis of activity situations.

By setting themselves the goal of comparing different moments in the activity situation using the concepts and methods they developed, researchers who were particularly interested in danger as a source of risk were able to detect and understand little apparent clues that could lead to an undesired outcome.

Comparisons can be made, either in relation to a desired situation taken as a key hypothesis and considered in vulnerable, or between two periods of time, or between two situations which have the same goals, but are geographically separated.

Thus, to enable a detailed analysis of the activity situation, it is important to characterize it.

2.2. Characterizing an activity situation

The characterization of a situation is carried out by an objective description of four parameters whose precision in the definition of their perimeter will impact the whole cindynics analysis. This **unavoidable phase of the study includes: a temporal and chronological horizon, a spatial and geographical horizon, a contextual description of the chosen horizon and the networks of actors involved in these horizons**, since their interactions generate complexity.

The choice of horizons and networks of actors is the result of a convention made by observers of the situation. The latter will have to justify their choices and reconsider them if the coherence and relevance of the cindynics approach require it[1].

– **Time horizon and chronology:**

It is a question of defining, in a space-time, the limits of the study of past facts and events (whatever their variety and complexity), and this for activity situations in progress. This space-time in which history was written can be several years long or reduced to one day. Moreover, the risk analysis of the Bhopal accident (see Chapter 8) shows that the global space-time can still be subdivided into several periods, because the relevance of the temporal delimitation of the analysis facilitates the subsequent identification of the vulnerability factors of the system being studied.

When a project is being studied, the future horizon is defined according to the life cycle envisaged for this project.

– **Spatial and geographic horizon:**

The choice of geographical areas will be dictated by:

– the characteristics of the environment: urbanization, ecological density, meteorology, natural aggressions;

– targets identified by the issues they represent: infrastructures, populations, systems important for safety and security;

---

1 See the section on the birth and activities of the working group in the Preface.

– available resources: human, technical, financial or organizational.

As an example, the spatial horizon taken into account must integrate characteristics (urbanization, etc.) in accordance with the chosen time horizon.

If the project is a work in progress, the features may be difficult to integrate.

– Description of the context:

“Before starting the design and implementation of the framework for management risk, it is important to evaluate and understand both the external and internal context of the organisation, since these can significantly influence the design of the framework” (ISO 31000:2018). Cindynics use this same method for hazards.

This description covers the characteristics resulting from the socio-economic, political, historical, regulatory, cultural and environmental conditions of the activity situation being studied. This step makes it possible to add new factors (that were not previously taken into account) to the current management practices and processes, which are focused on the triggering event.

Of course, the contextual elements must be consistent with the horizons chosen. For example, with reference to the space-time defined by the time horizon, the contextual elements in the regulatory data that were not, or are no longer in effect over the period cannot be retained.

– Networks of actors:

The networks include all of the actors identified during the description, whether they are from companies, administrations, political or institutional spheres, not forgetting the population.

These actors (the population, employees, managers, etc.) must also correspond to the chosen geographical dimension and context.

In conclusion, characterizing a situation of activities constitutes the grounding point of a risk analysis. The four parameters (Figure 2.1.) are interdependent and cannot be defined unilaterally. They must therefore integrate the limits of the other parameters, in the sense of defining the activity situation (Figure 2.2.).

The consistency of all parameters is essential to the quality of the cindynics process.

NOTE.– from experience, the group recommends using the “grounded theory” (see Preface) method to progressively build up all aspects of the characterization of an activity situation.

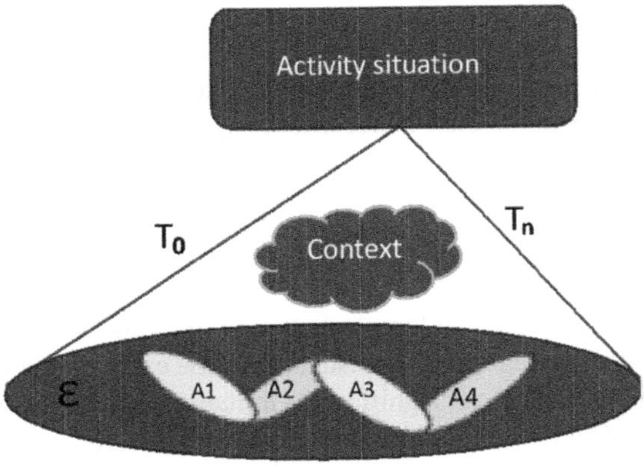

Figure 2.1. *T_0 to T_n: studied space-time = time horizon; E: space horizon; A1 to A4: actors taken into account in the activity situation. For a color version of this figure, see www.iste.co.uk/planchette/cindynics.zip*

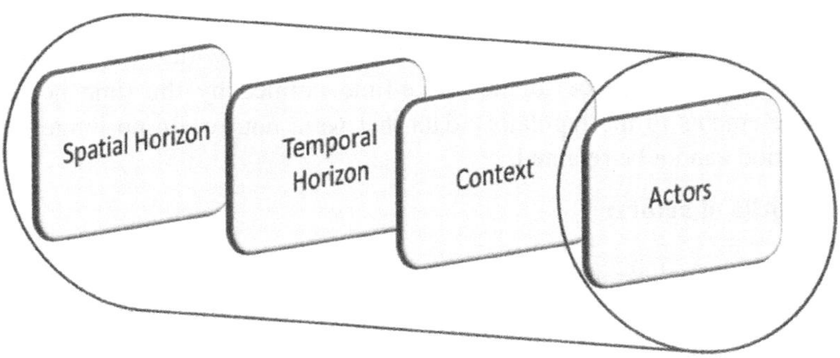

Figure 2.2. *Defining an activity situation. For a color version of this figure, see www.iste.co.uk/planchette/cindynics.zip*

2.3. Qualifying a dangerous situation within an activity situation

We have already mentioned the fact that the inevitable evolutions generate as many potential weaknesses as strengths within situations of activity (see Chapter 1, section 1.1), and that the cindynics approach focuses more particularly on weaknesses that are considered risk sources.

These fragilities lead to the notion of a dangerous situation within an activity situation.

2.3.1. *Notion of a dangerous situation*

A situation can only be considered dangerous when one or more issues – objective(s) or vulnerable target(s) – are exposed, voluntarily or involuntarily, to one or more risk sources (whether these are already part of a catalog or are in an unobservable form).

This configuration then creates a vulnerability of all or part of the organization and predisposes it to the occurrence of an accident, or even a disaster, if a random event triggers the materialization of risk sources.

An example taken from a paper presented at the 20th congress on risk management and dependability [DUB 16] shows that vulnerability is an important predisposing factor to a catastrophic situation:

> In terms of cyber threats, most of the time the offensive techniques used by hackers also exploit weaknesses in three facets of the complex socio-technical system: a technical element, such as a so-called "zero-day" flaw, human error (opening an attachment), organizational vulnerabilities, such as a lack of staff awareness [...].

All of these elements represent risk sources which, although may be considered perceptible, do not focus sufficient attention. Hackers exploit them and turn them into disasters that are reported in the press.

Figure 2.3 explains the process of constructing a dangerous situation and the impact of a random event that triggers a disaster.

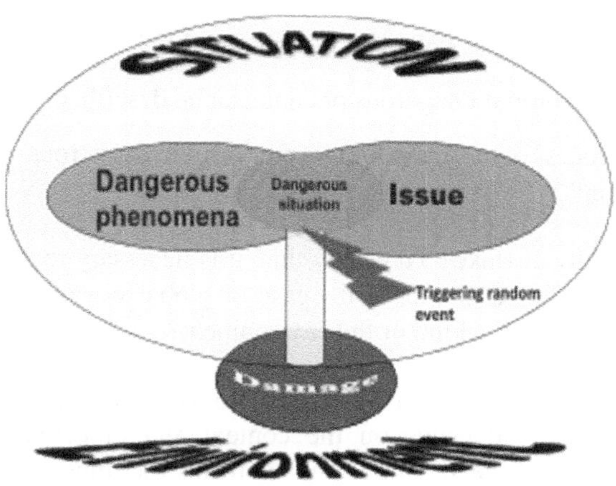

Figure 2.3. *Representation of a dangerous situation. For a color version of this figure, see www.iste.co.uk/planchette/cindynics.zip*

This representation highlights the fact that in order to cause damage regarding an issue, the following four conditions must be met:

– the presence of a dangerous phenomenon. Remember that a dangerous phenomenon corresponds to a capacity of release energy or substance representing a danger. Let us note that any phenomenon has, in an intrinsic way, capacities to damage an issue;

– the existence of an issue where the term “issue” covers elements of a material (goods, infrastructure, etc.), immaterial (image, intellectual resources, etc.) and environmental and people-related nature;

– a state of vulnerability of the issue, the fragility of which will promote the appearance of more or less disastrous consequences. Otherwise, its resilience will help the organization to absorb a disruption and subsequently regain its functions;

– the irruption of a random event, a hazard, a term often used by geographers. The hazard is a natural or technological phenomenon whose occurrence is more or less probable.

At this stage, it is useful to note that traditional management approaches focus particularly on the two characteristics of the hazard (likelihood of a random event, and seriousness of the consequences), often failing to address the vulnerability of the issue. And yet, this subject is as crucial as trying to characterize the hazard. Cindynics pay particular attention to this when characterizing activity situations (see Chapter 1, section 1.2).

Figure 2.3 therefore raises several questions:

– What are the phenomena that really represent risk sources for an issue?

– What is the real issue at stake and its vulnerability?

– How can the nature of dangerous phenomena be described?

– By what process can a random event turn a dangerous situation into an accident or disaster?

But what is really at stake? To clarify this, it is necessary to abandon prejudices or habitual reflexes. For example, in the case of global warming, which represents the threatened issue – humankind or the environment?

To answer this question, it is necessary to consider that the nature of the dangerous phenomenon depends on the context and objectives of each of the stakeholders involved in the situation. This confirms the importance of the study of activity situations (see section 2.2).

The answers to the last two points are found in the very essence of the cindynics approach, because "the science of danger is about researching the nature of the danger" [KER 07, p. 45], as well as the process that transforms the dangerous situation into an accident or crisis [KER 07, pp. 37–39] in order to improve prevention.

2.3.2. *Qualifying the dangerousness of a situation*

To qualify a situation (see Chapter 1, section 1.2) is to assign it attributes, qualifiers or aspects, essential means of observation allowing a **description** of this situation, where observable dangerous phenomena coexist with others not easily detectable.

The choice of qualifiers has already been mentioned. It is based on the characteristics of the organizations (data, models, rules, values and purposes) to constitute a qualification grid (see Chapter 1, section 1.3) or "hyperspace of danger".

The qualification of a dangerous situation is carried out by assimilating it to an Object Entity (recommendation of the RCM method – see Chapter 1, section 1.2), which facilitates its description by carrying out a **succession of views** (or glances) in regard to the tool "hyperspace of danger". Thus, it is possible to **describe what each of the networks of actors has at one's disposal in terms of data banks, models, goals and objectives, rules and norms, as well as values in order to evaluate the deviations from a situation taken as reference.**

This description is carried out either by **desk research** in the case of accident studies, or by **interviews with certain actors** of the organization in activity. In the latter case, the latter, whether external or internal, are part of the context of the situations and must be identified. **Here lies a real choice which shapes the volume of the studies,** because it would be pointless to think that all of the stakeholders of an organization could be interviewed. Therefore, it is important to seek out groups of stakeholders who are considered relevant and who can exert a significant influence on the situation. This choice is refined as the context of the situation is established.

The work of qualification also makes it possible to take into account the criteria of evaluation that the various actors have of the dangerousness of the situation. Indeed:

– they may be overconfident in their capacity for discernment, either leading to an attitude of excessive optimism (example of the Costa Brava disaster), or to an excess of anxiety leading to an excess of protection;

– the cultural systems in force also determine preconceived attitudes towards a risk source (denial of risk, language used, etc., for Bhopal) or overcoming oneself (example of the raiders in Antarctica).

Remember that the **description** of hazardous situations is based on the collection of data identified through a **questioning process.**

The questions, adapted to the situations studied, should be asked during interviews exploring the positioning of stakeholders at each of their levels (global, group and individual). The content of these interviews must then be carefully analyzed to avoid interpretation bias as much as possible, by using content analysis[2] techniques. Content analysis thus aims for a second interpretation of a message, to replace intuitive or instinctive interpretation with a constructed interpretation (Chapter 8, section 8.1.6).

Thus, the comparison of the glances between the situations (the reference one and the observed one) makes it possible to reveal the pathogenic elements contained in the situation. These may be, on the one hand, structural and organizational dysfunctions and insufficient knowledge at the level of each of the stakeholders qualified as **deficits** and, on the other hand, differences in points of view between stakeholders, referred to as **dissonances**. It is therefore necessary to identify and list them.

However, although the aim is to detect dangerous situations, the results relating to dissonances should in no case only be considered as new risk sources. These dissonances can also present advantages and reveal opportunities, just like bacteria, some of which can be very useful to humans. Moreover, it would be to deprive ourselves of the study of the positive dimension that these discrepancies may represent.

2 Content analysis is considered as "a set of techniques for analyzing communications aiming, through systematic and objective procedures for describing the content of statements, to obtain indicators (quantitative or not), allowing the inference of knowledge relating to the conditions of production/reception (inferred variables) of these statements" [BAR 77].

3

The Usefulness of Cindynics Tools

3.1. Qualification grid for risk sources that are not easily identifiable

Faced with the nature of risk sources that are less easily grasped, cindynics has adapted the method of relativized conceptualization (MRC) in order to describe them. Thus, a qualification grid could be obtained, based on the five qualifiers of the hyperspace of danger (see Chapter 1, section 1.3). These qualifiers, the most commonly used representation of which is presented in Figure 3.1 with the term "hazard hyperspace", allow a relativized description of these other types of risk sources to be obtained. This grid is therefore used as a qualification tool.

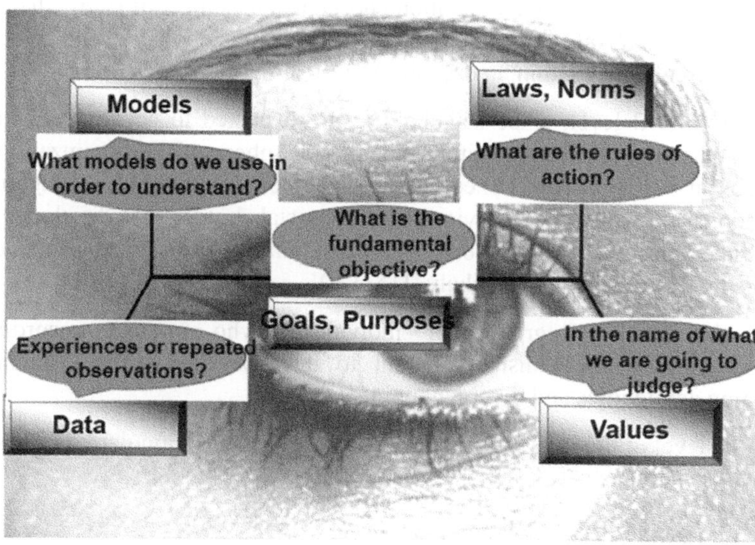

Figure 3.1. *Hazard hyperspace. For a color version of this figure, see www.iste.co.uk/planchette/cindynics.zip*

The meaning of the axes of the "hyperspace of danger" qualification grid are as follows:

Axis (1): dimension of facts, data and statistics. It answers the questions: what are the facts? What do the actors know?

Axis (2): dimension of representations and models developed from facts. It answers the questions: which models are the different networks of actors based on in order to understand and act? What representations do they make of them?

Axis (3): dimension of the purposes (social translation of the organization and its networks of actors). It answers the questions: what is the *raison d'être*? What is the fundamental objective?

Axis (4): the dimension of laws, norms, rules and standards, mandatory or free. It answers the questions: what are the rules, those imposed on the actors and those that the actors impose on themselves?

Axis (5): value systems dimension. It answers the questions: in the name of what to form an opinion, what ethical and moral references etc. guide the actors in their actions?

3.2. Describing this type of risk source

Based on this qualification grid (danger hyperspace), it is possible to describe the potential fragility of the activity situation chosen as the subject of study, thanks to a succession of views[1] on the behavior of the selected stakeholders. The fruit of the views is to be collected within an organization:

– At the global level, since stakeholders are expected to act together towards concrete goals while operating within the cultural sphere of the organization. This exploration at the global level sheds light on the context (cultural, managerial and organizational) in which the behavior of different stakeholders has developed, since the characteristics of the situations in which humans are placed shape certain behaviors by making them more likely [DAN 10].

– At the level of groups of individuals who interact according to psycho-sociological mechanisms inherent to their proximity, these interactions highlighting ambiguities and divergent points of view.

– At the level of the individual actor who possesses their singularities in terms of skills, representations, values and goals.

1 See Chapter 1, sections 1.2. and 1.4.

All the anomalies observed are, before analysis, considered as potential sources of risk considered not perceptible to our senses, because the behaviors, and especially the ingrained attitudes, are not always apparent, nor identical, even for similar situations.

3.2.1. *At the global organization level*

A thorough examination of accidents and disasters reveals that their origin comes from multiple causal chains distributed over time and combined in a pathogenic context [REA 13]. These causal chains therefore also include the managerial decisions that create the conditions for their triggering. These decisions are systemic, structural and organizational in nature. This pathogenic context is called **"Cindynics Systemic Deficits" (CSD).** A typology of 10 CSDs has been empirically established on the basis of numerous investigations carried out on a series of all types of accidents (see the section entitled "A bit of history" in the Preface. They have been classified into three main categories: **cultural (4)**, **organizational (2)** and **managerial (4)**.

The various topics presented at the 12th seminar, "Feedback on Industrial Accidents", organized by BARPI[2], which took place in Lyon in 2017, illustrate the concept of CSDs [PNR 17].

In addition, the report by Dr. Naoko Shimazu presented following the catastrophe in Fukushima insists on the different types of cultural deficits:

> It went on to state that "this was a disaster 'made in Japan'. Its fundamental causes are to be found in the ingrained conventions of Japanese culture: our reflexive obedience; our reluctance to question authority; our devotion to 'sticking with the programme'; and our 'insularity'" [SHI 12].

Cultural CSDs

These reflect certain pathogenic characteristics of the culture and therefore of the overall behavior of an organism. When an organism is immersed in a culture of success, when everything seems simple to it and when outside contestation is poorly accepted or when withdrawal is facilitated by the idea of not needing help, all these behaviors feed into "cultures" of wearing blinkers and of blindness conducive to ignoring risk sources.

---

2 Bureau d'Analyse des risques et Pollutions Industrielles (Office of Risk Analysis and Industrial Pollution).

To illustrate all the components of this type of deficit, let us present the case of the accident on the Deepwater Horizon (see Chapter 5) drilling platform.

Different sources of information mention (see Appendix 11) that:

– “The crew of Deepwater Horizon has the best safety record in the entire region and is admired by all. It’s a great confidence booster” [MAG 17]. This assurance, this certainty of success, is like a form of omnipotence that gives the illusion that everything is under control. This particularly dangerous behavior completely eradicates the feeling of being vulnerable to the peril of the security of the actors involved in the situation. The term “cindynics” is used to designate that a behavior **feels infallible**.

– Throughout the chronology, a form of laxity has gradually set in, dangerously lowering the barriers of vigilance. For example, the British Petroleum team decided not to carry out any further assessment of the performance of the cement at the bottom of a well. This approach to risk management, which seeks simplification by rejecting the idea of complexity or even innovation, is assimilated to a culture of **oversimplification**.

– The certainty of being the best, leading to withdrawal and therefore a lack of vigilance; the feeling of being the market leader, which saves the time spent by others to go elsewhere for information, leads to many forms of carelessness. On the Deep Horizon platform, people ignored or pretended to not notice accidents on other platforms. In this case, it was an example of **no-communication attitude**.

– “We detect several communication errors between the actors, particularly due to representation errors: indeed, BP was able to assume that Halliburton had carried out all the tests and the absence of other indications meant that the tests were good (which was not the case since they were not all carried out)” [MAG 17]. In addition, “Halliburton did not sufficiently inform BP and Transocean about the lack of sealing of the cement”. These examples highlight a **lack of communication** between partners.

Another example of a lack of communication concerns the event that took place in Gonfreville-L’Orcher on October 17, 2015[3]: a 100 m long by 4 m high cloud of ethylene was emitted from a chemical site. There was a risk of explosion of an unconfined cloud of vapors, yet the operator did not trigger its Internal Operation Plan (IOP) and the authorities were not notified until two days later.

Organizational CSDs

Resignation in the face of organizational tasks is a fairly widespread evil that explains many dysfunctions and disasters:

3 https://www.aria.developpement-durable.gouv.fr/fiche_detaillee/47277/.

– A few companies prefer to take action to save time and money, to the detriment of risk source analysis. This can be illustrated by the accident on an installation involving the ionization depolymerization of polytetrafluoroethylene (PTFE) [NIC 12] delivery times. In cindynics, these types of choices are translated by the **priority given to protection**.

– Analysis of the Mont Blanc tunnel accident shows that the responsibilities between France and Italy were poorly defined from the project design stage. Moreover, the creation of the originally planned Joint Tunnel Administration Committee never saw the light of day. Only an amendment specified that the tunnel would be operated by two concessionaires, one French, the other Italian. Many other examples can illustrate this context characterized by a **dilution of accountabilities**.

Managerial CSDs

Mixing with risk sources on a daily basis within an organization creates habits, even blindness, often translated by the belief that "accidents only happen to other people". As a result, the body is not prepared to face the occurrence of a dreaded event. This "recklessness" has been the cause of several accidents and disasters. D. Vaughan in 1996 refers to it in his study on the causes of the 1986 Challenger explosion using the term "normalization of deviance":

– For example, two identical accidents occurred 20 years apart (1989 and 2009) in a plastics manufacturing plant. Similar cases may occur or have occurred in other plants, such as the repeated explosion of a sulfuric acid tank during hot-spot work, with significant human and financial consequences. These examples indicate a **lack of a feedback policy**.

– Other categories of accidents, such as those that occurred in January 2013 at a chemical plant in Rouen, France and on April 3, 2016 at a company transporting hazardous materials (characterized by a fire followed by the explosion of tanker trucks), show another facet of this unpreparedness in the face of risk sources.

– In the first case, the prolonged rejection of mercaptan, due in particular to the transfer – 16 years earlier – of a tank accompanied by equipment not useful to the current process, shows that this transfer was carried out without a modification management procedure.

– The second case reveals that, despite compliance with regulations, the studies do not include the notion of a risk source related to the parking of tanker trucks. We can also identify in this section, the absence of procedure, written instructions or traceability of the modifications carried out.

It is also important to note the numerous dysfunctions generated by **interface conflicts** between units, services, subcontracting and so on. Indeed, organizations are seeking to develop their reaction capacities, thus involving more frequent interactions both transversally and vertically and even inter-organizationally. These interfaces constitute points of fragility and tension between actors. The accident on the Queen Mary (see Chapter 8, section 8.2) gangway illustrates these aspects of criticality:

– These examples show a no risk-management methodology (interdisciplinary and multidisciplinary).

– The lack of staff training is unfortunately a common practice for some. The example of the Bhopal disaster highlights this. Operators were caught off guard and made serious mistakes. No training for risk management and safety personal falls into the category of **major deficits**.

– Another deficit category is related to the unpreparedness of teams that have to face a crisis situation. For example, the Finnish example of the pollution of a river by a nickel sulfate spill in July 2014 led to a poorly managed crisis situation. Indeed, "the local environmental authorities, informed one day after the leak had stopped, were not aware of the extent of the spill. The authorities also had difficulty finding experts during the summer period. The extent of the spill was revealed by the press the next day. At the end of the week, when the high death rate of mussels was discovered, the media put strong pressure on the authorities. The accident took on a national dimension. The authorities were criticized for not having communicated early enough and for not having warned of the consequences"[4].

– Since then, the regional environmental protection agency concerned has set up a crisis management procedure, accident simulation exercises and an on-call duty role. In this case, the situation demonstrated a **lack of preparation for crisis management**.

This empirical approach, established according to a list of 10 CSDs (see Appendix 5 for a summary table) and based on the first elements of an organization's description, makes it possible to highlight the general reasons that already generate vulnerability in the situation being studied. It should be noted that this list can be enriched by other deficits drawn from new feedback.

As an example, students – at the instigation of René Baeli, then director of the École d'Ingénieurs de Marseille (Marseille School of Engineering) – studied the CSDs corresponding to 16 historical, even legendary, facts. Over several years, he was able to build up a database and gather these historical studies into a book entitled *Les Archéocindyniques* which applies the principles of cindynics to 16 events that marked French history.

4 Base Aria no. 48764 – https://www.aria.developpement-durable.gouv.fr/fiche_detaillee/48764/.

Five of the best-known cases are presented in Appendix 6.

3.2.2. *At the level of stakeholder groups*

The term **"dissonance"** is used to define communication problems between stakeholders. This term covers all vagueness, ambiguity, differences or divergences of points of view, purposes and so on, between actors or groups of actors. In an organization, purposes are rarely appreciated in the same way by top management, risk analysts, trade unions, staff and so on.

Dissonances are potential risk sources. The identification of dissonances is an unavoidable means to complete the risk assessment of activity situations or to identify the transformation operators able to lead an organism in a dangerous zone.

Comparative analyses of the hyperspaces of each of the stakeholders involved in the same activity situation (see Chapter 2, section 2.2, Figure 2.2) will allow the identification of dissonances.

However, if at first glance, dissonance produces a risk source (cindynogen), it can also be reconsidered positively. In this case, dissonance is also a tool for assessing whether the situation creates an opportunity to reduce its propensity to move towards an undesired transformation.

NOTE.– This concept has been used to approach systemic family therapy that seeks to bring about dissonance [FOU 97]. It is currently being researched in conflict resolution in international relations [COH 13].

3.2.3. *At the level of the individual actor*

Here, the pathogenic context is referred to as a **deficit**. It is a question of understanding the individual deficiencies, inconsistencies and blurred choices and decisions of an actor from the qualification grid. This type of deficit is partly similar to the concept of human error accomplished by the final operator. With this difference, cindynics is not limited to operators.

At the level of the individual actor, **four types of deficits** are distinguished:

– When it is a question of the absence of one or more elements on a hyperspace qualifier, this deficit is called a **gap**. For example, some experts may eliminate the data for a cause of a product explosion if they are unable to reproduce this in the laboratory and if no explosion model exists.

– In the case of inconsistencies between two qualifiers of the same actor, these deficits are qualified as **disjunctions**, following the example of the definition of the word according to the Cambridge dictionary: "a difference or lack of connection between two things" [CAM 21]. For example, a disjunction exists if a humanitarian organization uses processes that lead to wasting donor money or if they use it for personal purposes. Or if a manager issues a note prescribing staff to wear safety shoes in a workshop and does not react during his visit when he sees that an operator does not take this requirement into account. In both cases, there is a disjunction between compliance with the rules and the facts observed.

– As for the lack of hierarchy or priority in each of the five dimensions, which leads to vagueness (too many rules or non-hierarchical goals for the same person in charge, for example), it is then a question of **degeneration**, in the sense that priorities are absent or imprecise.

When an actor represents a situation in their own way and cannot or will not change it, or when they are used to a process that they repeat, no matter the situation's development (i.e., if they have a fixed idea about how to complete a task; if they are a "habitual offender"), then there is the presence of a **blockage**, in the sense that there is behavior characterized by a refusal, and an inability to react to a situation.

The results obtained by the comparative study of the information thus identified and the information desired upstream by the principals will enable the identification of the transformation operators to be taken into account to convert the situation into an opportunity or to anticipate preventive actions to reduce sources of risk. The implementation of a prevention plan must respond to an unfavorable evolution of the situation under study.

4

Reducing Risk Sources

"The normalization of the gap between the observed and the desired situation is to consider the degraded situation unfortunately becomes a normal situation". This remark refers to the concept of normalization of deviance established by Diane Vaughan [VAU 96]. The standardization of deviance corresponds to the acceptance of a risk source that may be at the origin of a risk whose occurrence is thought to be so unlikely that designers, engineers and operators tend to underestimate its materialization in spite of possible dramatic consequences.

It is therefore imperative to act before the "normalization of deviance" takes hold.

Therefore, once the new risk sources have been identified, it is advisable to list the CSDs identified, grouping them into two matrices: one for deficits (Table 4.1) and another for dissonances (Table 4.2). The tables deal with the Métalleurop[1] accident as an example.

Table 4.1 includes the following elements:

– *Operators* – **gaps** in the models: their eventuality is based on empirical conduct and does not integrate physical and thermodynamic phenomena.

– *Management* – **gaps** in the data: management is only interested in the volume of production and the level of pollution. They are not interested in safety aspects.

1 IMdR Training Day on March 26, 2013.

Stakeholders	Purpose	Values	Rules and norms	Facts and data	Models
Workshop team		B	G	G Disj	G
Management General management	Disj	Dge	Dge	G	G
Staff representatives	Disj	B	Dge	G	Dge
Administration Inspection of classified installations	Dge	Dge	Dge	G	G
Equipment manager Technical service	Disj		G	G	Dge
Documentation managers Organization, Methods, Quality	Dge	Disj	Dge	G	G
Cindynogenic situations	Contradictions in the objectives	Strong values leading to blockages	Few references to the norms and regulations	No references to data or measurements	No models (Empirism)

B - Blockage **Dge** - Degeneration **Disj** - Disjunction **G** - Gap

Table 4.1. *Sample deficit matrix*

Actors / Actors	Stakeholder 1	Stakeholder 2	Stakeholder 3	Stakeholder 4
Stakeholder 1 (SH1)		These 2 SHs do not share the same purposes	Different perceptions regarding values for these 2 SHs	Disagreement regarding measurement references
Stakeholder 2 (SH2)			Divergences in point of view regarding the interpretation of procedures	Inconsistency between these 2 SHs regarding data
Stakeholder 3 (SH3)				Disagreement regarding the interpretation of internal regulation
Stakeholder 4 (SH4)				

Table 4.2. *Example of a dissonance matrix*

It should be noted that other representations of deficits and dissonances can be chosen, as shown by the choices adopted for the examples of approaches (see Chapter 8).

With the help of these two matrices, it is possible to reduce the types of risk sources identified within the dangerous situation by applying an **action plan (intentional transformation operators)** capable of minimizing the possible consequences of the deficits and dissonances identified. These action plans can relate to a new formalization of the organization's goals and values, to the training and information of stakeholders, to the adaptation of the organization and to the adequacy of resources to the goals.

5

A Comparative View Between Dependability and Cindynics

5.1. Introduction

5.1.1. *Dependability*

Since the 1950s, dependability has proven its effectiveness through considerable progress in improving product reliability and availability. The models, methods and tools that have been developed have raised it to the level of **a science of failures**.

Since then, new constraints have emerged because of:

– the plurality of environments (technological, natural, human, social, legal and organizational) in which our technical systems evolve;

– interconnection loops characterizing the notions of networks in order to improve the security and availability of our systems.

Due to the very great complexity created, these constraints generate innumerable sources of danger and unexpected behaviors that current dependability methods and tools do not know how to model.

5.1.2. *The cindynics approach*

The approach has been developed since 1990, with the aim of detecting and understanding the nature of the danger in order to better protect against it. This approach presents the originality of analyzing the behavior of the actors internal and external to the system, evolving in their own contexts and also in interaction.

By bringing together all research in this field, **cindynics** has become the **science of danger**.

5.1.3. *Dependability and cindynics seem to ignore or even compete with each other*

Therefore, we ask the question: are the science of failure and the science of danger in conflict or are they complementary? To better answer this question, we will explore the two approaches in relation with the evolution of our systems towards complexity.

5.2. What is a complex system?

Let us first explain the notion of a system. It is a group placed in an *environment* and equipped with:

– a *purpose*, corresponding to an economic, social or human utility;

– an *organized structure*, composed of a large number of technical, human and natural entities;

– a *communication network*, allowing an infinite number of exchanges (energy, information, etc.), both internally and with the environment.

Complexity – whose characteristics will be developed in Appendix 2 – stems from the many interactions between entities, leading to unpredictable system behavior and thus compromising the achievement of expected goals.

Analysts, through feedback, have found that the lack of predictability regarding the behavior of complex systems generates responses that are different from those provided by linear modeling. Thus, the usual modeling methods and scientific tools currently available and proven for linear systems are not adapted to complex systems.

5.3. Dependability approach – its strengths and limitations

5.3.1. *The scope of dependability*

Dependability is an essentially technical approach focused on technological systems to minimize the risk of failures and their consequences. Applied to the system and all its components, the approach is based on the reproducibility of events using modeling laws derived from feedback and based on reliable databases.

The approach is aimed at each of the sub-systems contributing to the operation of the system without necessarily taking into account interdisciplinarity.

Humans are most often considered as components of the system, to which we seek to apply, by analogy with the other components, the laws of human failure. However, there have been advances in this field thanks to the pooling of behavioral data from the "work team".

5.3.2. *Description of the system and its components*

This description is a fundamental step in identifying all the subsystems and their components with their technical characteristics and their operating conditions or life profile (all the situations that the system may encounter during its life). To the supplier data (lifetime, failure rate and maintainability) are added those resulting from user feedback codified in regularly updated databases.

5.3.3. *Functional analysis*

This step specific to the dependability approach consists, according to the French standard, NF-X-50-150, in searching, characterizing, ordering, prioritizing and/or improving the functions of the system. These functions are those belonging to the product (hardware, software, processes and services expected by the user). There are two types of functional analyses: *external*, to express the expected need, and *internal*, to define the technical solutions to achieve each of the expected functions. At the end of this phase, a global vision of the system's functioning is available, which will allow the identification of the dangers that could disrupt the achievement of its objectives.

5.3.4. *Process hazard analysis*

This analysis (PHA) involves the identification of hazards defined as potentially dangerous situations that could cause damage to the system. Only hazards of a technical nature are identified here, whether they are internal to the system (component failures including human error) or external (fire, nuisance, natural origin, etc.).

5.3.5. *Technological choices*

This fundamental step, specific to dependability, determines the technical structures of the system so that it meets the expected functions. Then, each element of the system is examined in the light of the specific objectives expected of it in the functional analysis, knowing (the basic premise of dependability) that compliance with the objectives for each element of the system must contribute to the contractual objectives. This leads to choices that make it possible to retain technological elements of recognized reliability (see the databases resulting from feedback) and to eliminate, if possible, others with unknown or poor performance.

5.3.6. *Identification of failures – analyzing risks*

The system being well identified and characterized by its operating parameters, we can then proceed to an analysis of the consequences when components fail (FMECA (failure mode, effects, and criticality analysis) or similar) which will consist of:

– identifying the modes of failure in the constituent bodies selected;

– acting on the causes as early as possible (prevention);

– acting on the consequences likely to affect certain targets (protection);

– developing scenarios of the most likely failures with the most feared consequences, depending on the severity of the failure and the vulnerability of the affected targets.

This failure analysis is considered a form of risk analysis.

5.3.7. *Strengths and limitations of the approach*

The dependability approach provides essential elements for the technical shaping of a system's equipment. It is particularly well suited to the project phase of a system but, apart from feedback, it generally takes little account of changes and conditions that may affect the system during its operation. It leaves little or no room for human interaction with the system, which is a key element in the operation of complex systems.

As a result, in the risk scenarios identified, the desired exhaustiveness is not achieved.

5.4. The cindynics approach

5.4.1. *The cindynic situation and its scope*

As we have seen previously, cindynics focuses on the description of dangerous situations, by defining the risky situation and a list of actors or networks of actors involved in the situation, within spatial perimeters and temporal horizons (see Chapters 2 and 3).

This bias is fundamental, because from the outset it takes into consideration a system composed of elements in dynamic interaction in a given environment that is likely to evolve over time.

For each of the actors and networks identified, the description continues, actor by actor, through a grid comprising the five aspects of the hyperspace of danger.

In this way, hazard potentials can be described through deficit and dissonance matrices.

After having identified the main hazards (some of them eliminated, controlled or limited "at the source"), the potential targets will be identified as well as their vulnerability to the feared events.

This completed step enables a risk analysis to be carried out using conventional mapping methods, comparing the probabilities of occurrence of the feared events with the estimated consequences in terms of severity and in terms of the management of these events. The analysis of risk mapping enables the implementation of a hierarchical prevention and protection policy.

5.4.2. *Strengths and limitations of the approach*

This cindynics approach makes it possible to take into account:

– the totality of the technical elements of the system;

– the human resources necessary for the operation;

– organizational elements integrating human and technical elements;

– the environment of the system, be it physical, social, economic, political, historical or legal;

– a temporal dimension which implies regular updates;

– a documentary dimension that lists the entire technical, regulatory, normative and deontological context in which the system evolves.

Moreover, the definition of the system's objectives goes beyond the definition of the technical production objectives alone and also makes it possible to specify economic and security objectives, and even social objectives.

While the cindynics approach has the benefit of giving a leading role to the different actors concerned, their identification is not simple. For it is possible to integrate too many actors not concerned by the system or, conversely, not to identify "hidden" actors. Moreover, the adapted method, which focuses on the relationships between the actors, does not *a priori* use all the resources of functional analysis and the definition of technical structures. Also, the technical shaping is neglected, which does not contribute to the systems' technological design phase.

5.5. Conflict or complementarity of the two approaches

Dependability criticizes the complexity of cindynics in terms its approach as well as the difficulty in understanding its concepts. The cindynics conceptualization, on the contrary, criticizes dependability for its oversimplification in describing the potentiality of events, limiting the reasoning on risk to a two-dimensional space (probability and gravity), whereas cindynics details its reasoning from five dimensions. Moreover, human and organizational factors are far too much neglected by dependability, the complexity of these factors being assimilated only to the behaviors of technical components of the system.

In fact, dependability actively contributes to the definition of the structure, the technical shaping and the choice of components that will enable the realization of the best system corresponding to the availability objectives set by the ordering institutions. However, the basic principles are only based on technical approaches. While it is well known that organizations are governed by techniques, structures and rules, research has shown that they are also driven by the values and beliefs of their members. This is the source of the tensions between rules and values that often lead to paralysis. To a large extent, it is values and purposes that drive the behavior of individuals and organizations.

Thus, dependability cannot provide a sufficient response to the complexity of systems and in particular to their human and organizational dimension.

On the contrary, while the cindynics approach does not provide an answer to all the technical aspects taken care of by dependability, it integrates the different components (equipment, humans and organization) of a complex system to highlight the tensions, deficits and dissonances, which are also at the origin of risks that cannot be identified by dependability.

In summary, if we consider the life cycle of a system:

– the operational safety approach is particularly adapted to the design phase, to the definition of operating modes and to the monitoring of tests and commissioning, where it can contribute to technological choices ensuring optimal reliability of the system;

– the cindynics approach is more specifically adapted to the monitoring of the system's operation and, by studying the interplay of the different actors, allows the implementation of policies to reduce the tensions that may arise between the different actors of an organization and between the actors and the organization itself. The two approaches therefore have the same overall objective, that of reducing the various risks, whether technical, human or organizational.

5.6. Conclusion

While the two approaches have had different histories and come from different cultures, they are far from conflicting. On the contrary, **they are very complementary**.

To conduct them properly, they require the mobilization of diversified teams, technically specialized for dependability and multidisciplinary for the cindynics approach.

This is the major reason why the *Institut pour la maîtrise des risques* (French Institute for Risk Management) has integrated these two approaches while continuing to develop their specificities. This is because the IMdR (*Institut pour la maîtrise des risques*) has become convinced that their joint and judicious use is perfectly adapted to the study of complex systems.

6

Perspectives

Often misunderstood because of its new specific vocabulary that it adopted at the outset, and the change it caused in our habits of constrained thinking due to our Cartesian training, the cindynics approach encountered many difficulties in convincing us of the power brought by its concepts. Furthermore, because it was initially based on experience feedback (REX) from accidents and disasters, it was sometimes criticized because it was reduced to a mere tool for establishing and analyzing REX.

After a period of development, thanks to teaching in universities and large schools in France (Université Paris 1 Panthéon-Sorbonne, Faculté de Lyon 3, Écoles des Mines de Paris, Nancy and Saint-Etienne, Master's degree in Territory Management from the École Internationale des Sciences du Traitement de l'Information (EISTI)), the cindynics approach has experienced a certain decline.

However, since 2012, two Institutes, the *Institut pour la Maîtrise des Risques* (IMdR) and the *Institut Méditerranéen des Cindyniques* (IMC) have been breathing new life into it by promoting the interest and robustness of cindynics concepts. In view of its geographical location, the IMC gives priority to the study of the vulnerability of firms to wildfire and crisis management[1]. As for the IMdR, this institute runs periodic training courses so that the approach is accessible to a large majority of managers in both the industrial and local authority sectors (see, Appendix 13). This is the role of the working group and the specific reflection group which presents its main results here.

---

1 See reports on the website: https://www.cypres.org/default/sciences-du-danger.aspx?_lg=fr-FR.

In addition, the IMdR contributes to various studies and other working groups with the objective of promoting cindynics.

One example is the studies carried out with the French Association for the Prevention of Natural Disasters, *Association Française pour la Prévention des Catastrophes Naturelles* (AFPCN).

Between 2009 and 2012, these two Institutes set up a working group with the aim of proposing preventive actions capable of reducing the vulnerability and increasing the resilience of the networks (water, electricity, gas, etc.) that constitute the economic and social fabric of France. Indeed, several major climatic events in the last decade (Vaison-la Romaine floods, storm in 1999, marine submersion in 1997) have shown the vulnerability of these networks. In order to do this, the working group adopted the cindynics approach, a method that had never before been applied in natural risk studies. At the end of the work, a summary of their reflections was produced (see Appendix 12).

At the end of 2014, the IMdR also created a working and reflection group entitled "*Management des Risques, Cindyniques et Nouvelles Approches Systémiques dans le Secteur de la Santé*" (Risk Management, Cyndynics and New Systemic Approaches in the Health Sector). Indeed, as the health sector emphasizes, more than any other sector, the share of human and organizational factors in adverse events, it seemed important to develop new approaches that take into account the entirety and complexity of the healthcare activity. This group is therefore an opportunity to create exchanges and share experiences between the world of technology and the world of healthcare. This working and reflection has therefore set itself two objectives with the aim of:

– sharing:

- knowledge and understanding of new approaches to risk management (systemic, cindynics, HRO, use of simulation and training tools, etc);

- concrete experiences of risk management implemented in the healthcare sector, particularly the risks of integrating the patient and taking into account human and organizational factors;

– encouraging relationships between sectors.

As a result, the IMdR considers that the work carried out since 2012 opens up new prospects for the development of cindynics.

Indeed, by placing human organizations at the center of studies of risk sources, the cindynics approach appears well suited to the conduct of strategic studies. In this respect, it should be remembered that in the preface, emphasis was placed on the "cultural aspects and individual games" existing within organizations. By broadening the approach to "political, military games, [...]", these principles of the cindynics approach can also be adapted to describe the deficits and dissonances contained in conflict situations, and then to understand them, in order to reduce their scope and intensity. Such a process has already been cited as an example that has made it possible to bring the dissonances existing in a family system to light, as much as those contained in conflicts within international relations.

With regard to the study of crisis management, the French Society of Disaster Medicine, *Société Française de Médecine de Catastrophe* (SFMC), after having effectively developed a tactical process for the treatment of disaster victims, is making progress in its strategic approach in terms of crisis committees. The "conceptual contributions developed with cindynics and confirmed by the simulations carried out at the École des Mines de Nancy should make it possible to find optimal solutions for the operation of a crisis committee" [HAN 17]. Indeed, at the end of these simulations, the feedback from the experiments carried out in Nancy highlighted correlations between the conceptual tools proposed by cindynics (deficits and dissonances) and the results of the observers.

As for the field of standardization, the IMdR was able to integrate the cindynics tool during the work on the revision of the IEC/ISO 31010:2009 standard.

All of these perspectives are of great interest for the development of cindynics. However, it is imperative to continue the recognition of this approach by persevering with research work, in particular for:

– broadening the scope of use of the principle of describing dangerous situations to that of timely situations. Indeed, this approach to studying situations has the capacity to identify threats and weaknesses (deficits and dissonance), as well as opportunities and strengths that are not perceptible in the organization. We return to the idea developed previously concerning the propensity to create an advantage. The cindynics approach can perfectly complement [HUM 10] SWOT studies (Strengths, Weaknesses, Opportunities and Threats), a business strategy tool that helps determine the options available in a strategic area of activity;

– learning to better conduct systemic studies and their possible modeling, to understand the notions of emergence within complex systems, in order to better manage those that are not desired;

– establishing an appropriate selection of actors or stakeholders for their relevance;

– studying one or more questioning methods (see Chapter 2, section 2.3.2) to enable the stakeholders' positioning at each of their levels (global, group and individual) to be collected, as free of preconceptions as possible;

– specifying the methods for analyzing the content of the collection of interviews;

– better defining the use of the notion of propensity (see Chapter 2, section 2.1), its links or differences with the notions of probability for uncertain future events;

– improving the operational nature of the cindynics approach, in order to establish a hierarchy of risk sources to better assess the classification of preventive actions to be taken;

– developing industrial applications, particularly in the context of plant design and maintenance optimization.

Conclusion

The lessons learned from the work carried out within the framework of this "Cindynics for All Working and Reflection Groups" reveal that, by incorporating the human and the organization, which are at the center of the complexity of systems, the cindynics approach can integrate all the richness brought about by new scientific, philosophical and sociological advances in order to grasp the complementary potential of dangerousness that can exist within organisms.

Thus, in order to better **evaluate and deal with all risk sources, whether known or potential in activity situations, this approach helps to explain how dangerous situations are created**. This exploration is carried out by studying the inevitable evolutions that organizations undergo, their purposes, the interactions between members of work collectives, individual and inter-individual games and value systems, which, before now, have not all been used in the same approach.

A cindynics study is therefore not limited to the role of retrospective investigation, as post-disaster forms of feedback are. It **now allows** *a priori* **analyses of risk sources that are not easily perceptible**, which will enrich the risk analyses currently being carried out.

Methods and tools can now help to identify a complex system, discern deficits and dissonances within it, and identify areas of vulnerability with a view to reducing them. This involves making an organization's structures more resilient, either to find a new balance or to adapt to changing contexts in time and space. This resilience capacity limits a system's natural entropy towards disaster.

In addition, the cindynics approach makes it possible to structure **an organization's search for coherence** between its goals, its values, the rules it must know and follow and the data and models it must use.

Indeed, every organization sets qualitative and quantitative objectives to be achieved. But how do they ensure that the pursuit of these objectives is compatible with their values, their models and their rules? Some networks are tempted to use different means to reach the set objectives or even to inform their partners about them, etc. Sometimes it turns out that they are not wrong… provided that they have been able to evaluate the price to pay in the future.

This search for coherence is in line with the **image of homeostasis**, which designates the result obtained by bringing into play different types of regulation, ensuring equilibrium in living systems that are always dynamic.

The entire approach is therefore an invitation to re-examine the contributions of our current risk management approaches: because these approaches are generally based on the initiating event and are limited to the observation of human activities linked to the actions of the final operator leading to "active errors" [REA 13]. However, "while the notion of human error is (still) popular today, it still contains some limitations, because this crude term says nothing either about the pathogenic context factors that acted on the operator at the origin of the human error", Reason specifies [REA 13]. Having attempted to define a generic model of human error summarized in three main types (failures, lapses and errors), Reason finally rejected this limitation in his 2013 work, mentioning that "in future research, it will be necessary to cover organizational as well as individual failures [...] because it is precisely social and institutional factors that now represent the greatest threat to our safety". In addition, Reason notes that:

> [When] examining the human contribution to systems disasters, it is important to distinguish between two types of errors: **active errors**, whose effects are felt almost immediately, and **latent errors**, whose adverse consequences may remain dormant in the system for a long time and only manifest themselves by combining with other factors to breach system defenses (see [RAS 84]).

This evolution of thought does not consist of wanting to transfer the responsibility for human error from the operators to organization management, but in trying to make people realize that the understanding of importance of safety barriers, control operations, etc. is diluted in time during the slow transformations caused by changes in strategies, reorganizations, productivity research, staff renewal or reduction.

A recent example perfectly illustrates this notion of latent error, that of the explosion that occurred on Tuesday, August 4, 2020 in Beirut at 6:10 p.m. with the origin being a stock of 2750 tons of ammonium nitrate, an industrial chemical product commonly used, mainly as fertilizer, but also in the composition of explosives. But this dangerousness is well known, as it has already caused several industrial accidents, including the 2001 explosion of the AZF plant in Toulouse, southwestern France, where some 300 tons of ammonium nitrate exploded, killing 31 people. This product – seized from a Moldovan-flagged ship from Georgia – had been stored since 2013 in warehouses in the port of Beirut and kept without special surveillance. The carelessness and slow transformation caused this time bomb to turn – seven years later – into a live bomb reducing the city to ruins. The event(s) that caused the destabilization of the ammonium nitrate is (or are) not yet specifically identified.

Through its approach and the deployment of methods and tools, cindynics is a response to the research perspectives on social and institutional factors mentioned by J. Reason.

Moreover, it highlights the fact that "the greatest threat to our safety" also lies in the slow changes that business situations inevitably undergo, thus justifying its founding concept.

Examples of Approaches

E.1. The Bhopal disaster

E.1.1. *Context*

At the end of the 1970s, a factory producing pesticides from a volatile and highly toxic compound (Sevin and Temik) was set up on the outskirts of the city of Bhopal in the state of Madhya Pradesh, India. This plant was built to produce 5,000 tons of pesticides a year. The installation of this plant, a subsidiary of Union Carbide Corporation, was welcomed with open arms by the Indian authorities who advocate a policy of agricultural development and crop intensification through a "green revolution". Located 5 km from the city center, the plant attracted a rural population seeking jobs. The massive introduction of these workers was being carried out right up to the factory gates, without any urbanization plan. The population of the greater Bhopal area thus rose from 225,000 to around 700,000 over a period of nearly 20 years (1961–1980).

From the beginning of the 1980s, economic difficulties led to a reduction in demand, affecting production, which began to slow down and never reached the 5,000 tons planned. The factory thus becomes a loss-making venture. A savings plan was then launched: the workforce was reduced leaving only the maximum possible, trained staff were replaced by other less competent employees. In December 1984, the plant was idling and operations were reduced to maintenance.

E.1.2. *The disaster*

It was during maintenance operations at the plant, which had been shut down for a month, that the disaster occurred on December 2, 1984. At 8:00 p.m., maintenance technicians proceeded with a pressure cleaning of the pipes connected to a tank containing 42 tons of MIC (methyl isocyanate) used as a base for the production of pesticides (see Appendix 8). Noting that the water was not flowing as expected, they alerted the foreman. A change of maintenance team took place around 10:30 p.m. Around 11:00 p.m., a major leak of MIC was noticed, but the shift supervisor did not believe that an accident had occurred and ordered a "tea break". During this time, impurities from corrosion of the pipes flowed back into the tank and produced chemical reactions and an abnormal build-up of pressure. Around midnight the pressure in the tank was double the maximum working pressure. This was noted by the foreman. To remedy this, the procedure involved sending the toxic gases to a decontamination unit and then to a flare tower to be burned. However, the decontamination unit and the flare tower had been taken out of service to reduce costs. The gases then blew out of the closed valves during cleaning operations and could not be burned off either. As a result, the deadly gas spewed into the sky for two hours and, when the wind blew, it headed towards the station and the city center. The alarm siren was only switched on an hour later and stopped after a few minutes. It was only around 3:00 a.m. that the tank was emptied.

E.1.3. *The consequences*

The human toll was extremely high. There were nearly 3,000 deaths and 200,000 injured. In addition, nearly half of the pregnant women exposed within a radius of 1 km around the plant had miscarriages or stillborn babies. In the impacted area, four years after the disaster, the mortality rate was still 4% higher than the national rate.

However, there was chronic pollution prior to the accident in the soil and water which had been contaminated by plant effluents rich in heavy metals and toxic substances. In 1999, the soil still contained high quantities of mercury, chromium, copper, nickel and lead. In 2003, mercury levels were six million times higher than current standards. Chloroform, carbon tetrachloride and benzene were detected in well water south and southeast of the plant. Vegetables and fruit still contain high levels of these toxic compounds. This chronic pollution, aggravated by toxic discharges, has long affected the population.

Thousands of tons of toxic waste are still stored in poor conditions, contaminating the water in the public network. Some sources report the presence of 25,000 tons of solid waste on the site. One journalist claimed to have seen mercury in ponds, dumpsters filled with toxic substances and bags containing very dangerous chemicals in some of the hangars. He also pointed out that during the monsoon season, rainwater that seeps into the site becomes loaded with toxic compounds, polluting the wells where people come to drink[1].

At the economic level, in 1989, only the medical impacts were taken into account when putting the amount of financial compensation at $470 million (while compensation for the clean up of the site was estimated at more than $2 billion). Union Carbide, the owner of the plant, has never assumed responsibility for cleaning up of the site. Since 2001, Union Carbide's financial assets and profits have been taken over by the giant multinational Dow Chemical, which has always refused to take over responsibility for the Bhopal disaster.

E.1.4. *The various analyses of the disaster*

Documentary research found that during the judicial inquiry, two opposing theories were put forward to explain the cause:

– the cleaning of the pipe, which was hampered by technical impossibilities during the construction process;

– the deliberate and malicious introduction of water into the MIC tank by a disgruntled employee who had been denied a raise.

The version pointing to sabotage raised much controversy in India, especially since the author of this investigative report was commissioned by Union Carbide.

Diverging opinions of witnesses and experts and changes over time in good or bad faith testimony mean the truth may never be known.

Similarly, the way in which the water entered the tank still divides specialists today, in terms of how the chemical reactions could have led to this very high rise in pressure:

– an (unlikely) explanation related to the partial hydrolysis of the MIC with the attack of the tank walls following a succession of reactions;

– a (generally accepted) explanation of a sudden polymerization of MIC in the presence of impurities acting as a catalyst and resulting in a significant release of heat vaporizing a significant part of the MIC.

Subsequently, from a methodological point of view, several modes of causal explanation have been proposed by various experts.

---

1 https://www.aria.developpement-durable.gouv.fr/wp-content/files_mf/FD_7022Bhopalinde_1984_fr.pdf.

The uni-causal and linear approach according to Jeanne and Roger Kasperson (see Appendix 7)

More of an explanatory model than a model looking for the chain of multiple causes that led to the accident.

Grenouillet's multi-causal tree of causes approach (see Appendix 7)

This study is intended to show that an accident is the result of the sequence and interaction of several technical and human factors.

The systemic approach according to Pauchant and Mitroff

Drawing on the principles of system analysis, these two experts proposed a model that reveals some twenty contributing factors whose sometimes complex interactions they highlight (see Appendix 7).

The filter method approach according to J.-C. Le Coze et al.

This method makes it possible to represent the study of an accident in the form of filters or levels of information according to different levels (operational, management, etc.), which depend, on the one hand, on the quality and quantity of the information describing the accident and, on the other hand, on the expertise available to process this information (see Appendix 7).

E.1.5. *Critical analysis of these approaches*

These different approaches may reveal limitations if we want to take into account the complexity of interactions within such a system. The filter method integrates a set of interactions, but these interactions remain centered on the Bhopal production plant. The other actors (American and Indian decision-makers, etc.) are not taken into account.

Moreover, the temporality is concentrated practically around the day of the accident.

The cindynics approach will broaden the scope of the investigation to examine all the vulnerability factors that contributed to the magnitude of the disaster.

E.1.6. *"Bhopal" with a cindynics approach*

In view of the available data, the IMdR's working and reflection group "*Cindyniques à la portée de tous*" (Cindynics that everyone can understand) based its work on the study of the characteristics of the **evolution of the situation up to the accident** in order to identify all the sources of potential risks. For, in addition to the sources of risks that are commonly known, this approach makes accessible the dangers linked to the interactions between all the networks of actors acting within the situation, to their differences in culture, beliefs, and so on – dangers that are not easily understood.

In order to better understand the evolution of the situation, it quickly became apparent that it was essential to conduct documentary research **covering the entire lifecycle of the pesticide production plant**, from its design to the day of the disaster, using various works on the Bhopal disaster (see Appendix 8).

The work was completed in several stages through analyses produced in subgroups, each subgroup presenting its results in turn. The process **began with a content analysis of the sentences collected through the documentary research** (sentence-by-sentence content analysis, breaking down sentences containing several ideas so that each sub-sentence contained only one idea). It then continued with the development of a table grouping together themes with certain "resemblance", then by a **reclassification of the different information according to the dimensions of the cindynics hyperspace** (purposes, values, data, models and rules) (see Chapter 3, section 3.1) in order to then decline the deficits and dissonances.

The group was thus able to characterize the situation (see Chapter 2, section 2.2) by **progressively refining the temporal, chronological, spatial and geographical horizons**, then by broadening the context (see Chapter 2, section 2.2), through the technical, meteorological, socio-economic, political and regulatory contributions contained in the additional information collected on Bhopal (see Appendix 8), and **finally by choosing a group of actors deemed relevant** among all those cited by this same information.

E.1.7. *The activity situation*

The temporal and chronological horizon: in order to understand the genesis of the disaster, the study could not be limited to the moment of the accident. A more in-depth study of the different stages in the company's development, from 1957 to 1984, provided elements for understanding the causes of the accident, the management of the event and the consideration of the victims.

Analysis of the content of the information enabled the working group to define three acts in the Bhopal tragedy.

– **Act 1 concerns the prologue for the years (1957–1980)** which marked the birth of the project to build the plant.

– **Act 2 (1980–1984)** concerns the operation of the plant.

– **Act 3 (the night of December 2–3, 1984)** focuses on the accident with its terrible consequences.

The spatial and geographical horizon was delineated progressively by the exploration of the areas at the different moments of the acts.

It includes the following players: Union Carbide Limited (USA) and UC India, India, the State of Madhya Pradesh and the territory of the city of Bhopal.

As for the *context*, the elements are based on additional information relating to:

– Union Carbide in the USA;

– India from 1950 to 1984;

– the state of Madhya Pradesh where the factory was built;

– the subsidiary Union Carbide India Limited;

– the Carbide factory in the state of Madhya Pradesh;

– housing developments;

– the products used in the manufacture of the pesticide;

– the attitudes of those in charge.

It should also be noted that during this night, the wind was blowing towards the slums and the city center.

The choice of *actor networks* was made according to the temporal, spatial and contextual criteria retained above in the situation studied.

The networks of actors selected for this presentation and according to the defined horizons are as follows:

1) 1957–1980: India, the state of Madhya Pradesh and Union Carbide Limited (USA);

2) 1980–1984: UC India, employees;

3) 3 and 4 October, 1984: UC Bhopal, City of Bhopal, population, employees, relief.

The dangers not identified according to the usual methods of analysis were identified according to the three levels included in the cindynics approach: that of the general causes inherent to the organizations (systemic deficits); that of the insufficiencies at the level of each actor network (deficits); and that of the divergences between several networks (dissonances). The qualification of these dangers was established at the global level of the organization as well as on the five axes of the hyperspaces of danger associated with the networks of actors: purposes, values, rules, models and data.

E.1.8. *Cindynics Systemic Deficits (CSDs)*

Which CSDs have been highlighted? With the help of the table (Appendix 5), we can mention some important ones (see Appendix 8):

CSD 1 – *culture of infallibility*: the feeling of near infallibility of Union Carbide's processes in the United States which were then transposed to India without change.

CSD 2 – *culture of oversimplification*: the experience of a financier, to run the plant without the assistance of a hazardous products specialist, was deemed sufficient.

CSD 3 – *communication culture*: no information was given to the plant's medical staff, neither on the toxicity of Sevin, nor on the pathology of accidents that may be caused by methyl isocyanate.

CSD 4 – *navel-gazing culture*: no research on processes carried out in other countries, for example for the storage of MIC.

CSD 5 – *priority given to production over safety aspects*: a safety audit carried out by UC in Bhopal in 1982 revealed more than a hundred anomalies that were not dealt with, water pollution leading to the death of several cows, the factory compensated the owners but did not disclose the results of water analysis.

CSD 6 – *dilution of responsibilities*: the Government of Madhya Pradesh refused to validate a comprehensive evacuation plan proposed by the plant manager in the run-up to elections.

CSD 7 – *lack of feedback*: scant analysis of precursory incidents and no plans for improvement after MIC leakage findings.

CSD 8 – *lack of procedure with a cindynic dimension*: no comprehensible instructions for staff speaking only Hindi.

CSD 9 – *lack or inadequacy of risk management knowledge*: as an example, the director, seeing the plant as being essentially shut down, had halted the main process-specific safety systems (including MIC refrigeration and flaring); the workshop manager did not know the maintenance rules, having newly arrived. He ordered a tea break instead of inquiring about the importance of the reports.

CSD 10 – *lack of preparedness for crisis situations*: doctors on call at the hospital saw a flood of patients who died on arrival. They learned of the accident at the Carbide factory but did not identify the dangerous substance – absence of a crisis plan.

E.1.9. *The hyperspaces of danger (HED) realized*

Each hazard hyperspace fits into the activity situation defined for the study.

Our first examples give the result of the analysis on an identical temporal horizon for two identified network actors.

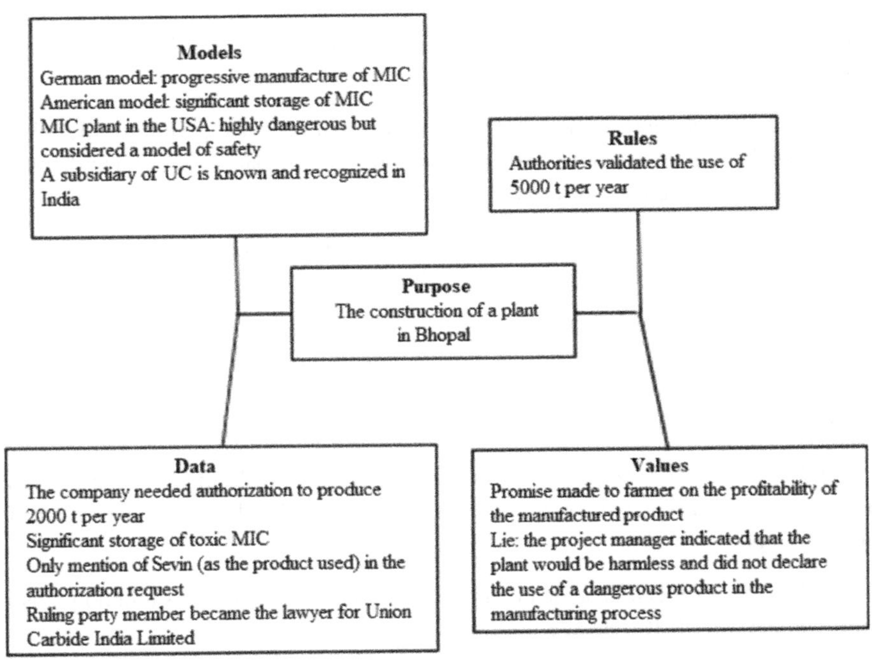

Table E.1. *Hyperspace of the actor Union Carbide (UC). Temporal horizon: 1957–1980*

An analysis of Union Carbide's HED from 1957 to 1980 revealed the following *deficits:*

– discrepancy between the German model (manufacture of MIC as it is produced) and the American model (large storage);

– failure to provide information to authorities: value axis *gap*;

– the plant manufactured 2000 t of Sevin for an authorization of 5000 t, therefore oversizing of need: *disjunction* between data and rules;

– *disjunction* between values and models: lie/transparency about the safety of the plant.

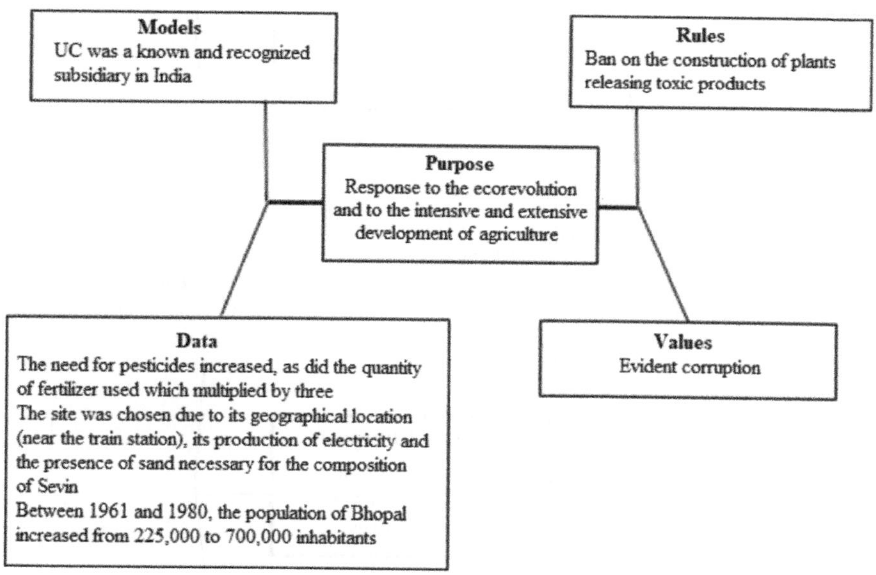

Table E.2. *Actor: "India and Authorities". Temporal horizon: 1957–1980*

The HED of the "India and Authorities" system over the period 1957–1980 highlights *deficits*:

– *gap* on the values axis: presence of corruption;

– *degeneration* on the data axis: no correlation or hierarchy between the geographical location of the site and the evolution of the population near this site.

The consideration of these first two HEDs immediately shows *dissonances*:

– The "India and Authorities" HED indicates on the rules axis: prohibition of chemical plants, while the HED of UC specifies on the data axis: significant storage of MIC.

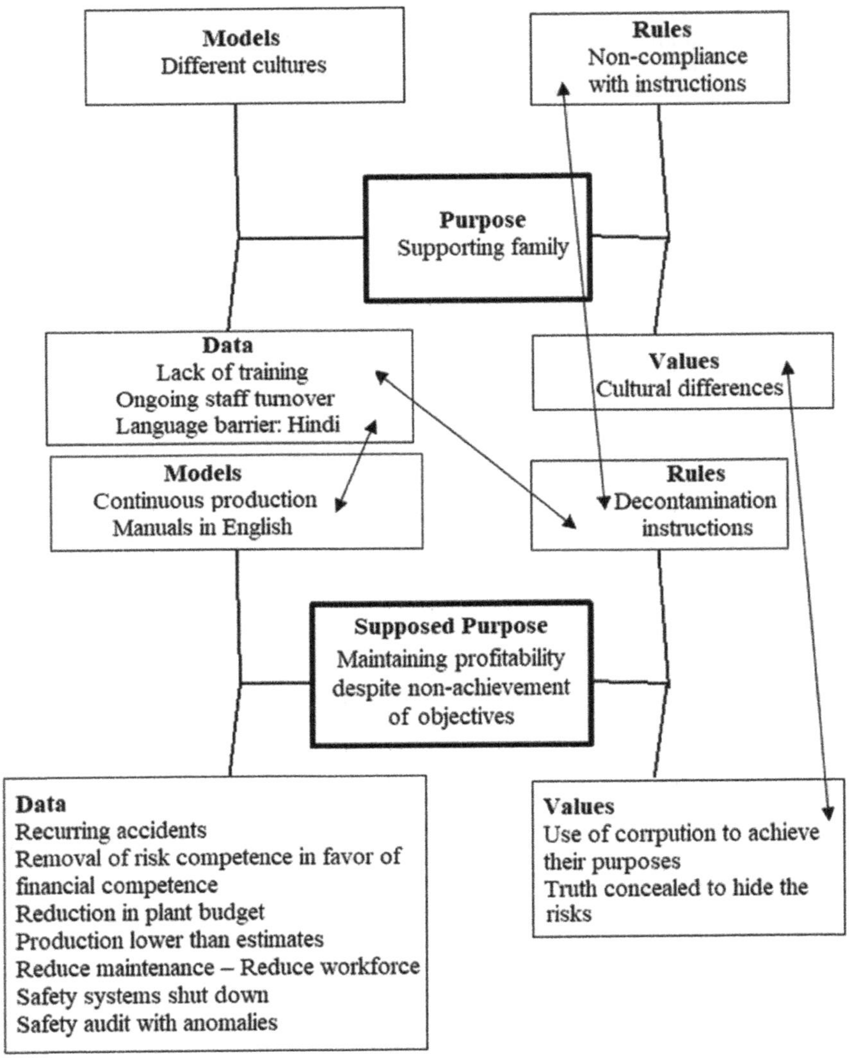

Table E.3. *Actors, employees and 8.1.4. UC India, stakeholder, period 1980–1984*

In the second example, we extracted the HED deficits of two actors over the 1980–1984 time horizon then identified the dissonances by comparing the two HEDs.

From the HED of employees, we have identified the following deficits:

– *gaps* on the rules axis: non-compliance with instructions;

– *degeneration* on the data axis: staff turnover/lack of knowledge due to lack of training/language barrier.

Similarly, Union Carbide's HED allows us to extract the deficits:

– *disjunctions* between data (staff cuts, safety shut downs, budget cuts) and models (continuous production);

– *blockage* between models and data: although the plant maintained continuous production, staff numbers were reduced and safety systems shut down;

– many accidents not taken into account: lack of feedback while evaluation tools were put in place.

Comparison of the two HEDs: employees and UC India between 1980 and 1984:

Dissonances appear immediately between the two hyperspaces studied:

– employees spoke Hindi and UC's instructions were in English;

– staff turnover was constant, training was lacking when employees needed to know safety instructions (especially decontamination instructions).

The graphical exploitation of these two HEDs illustrates these dissonances definitively.

The last time horizon considers the date of the accident and the observation focuses in particular on Union Carbide's HED.

New *deficits* appear in the analysis:

– *productivity criterion dominance* on the purposes axis: dismantling, transfer of production in breach of safety rules;

– *no information* on the values axis of MIC toxicity data from the physician;

– on the model axis, what is observed does not enter into the thought model: leakage of MIC ignored by the shift supervisor.

In this last horizon studied, technical elements common to other cindynics methods appear.

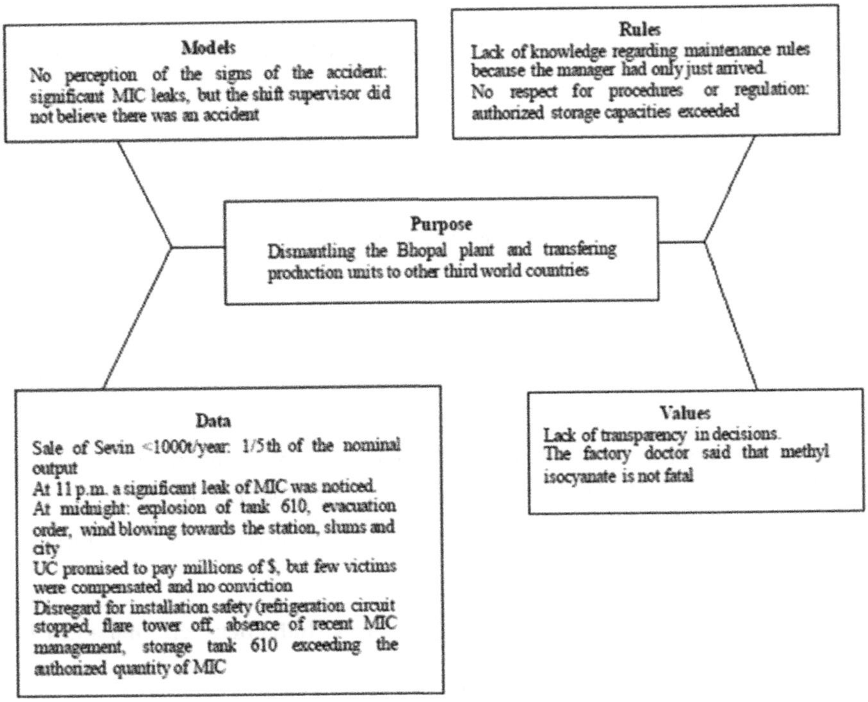

Table E.4. *Actors: Union Carbide – October 3 and 4, 1984*

The exploitation of the HED of UC (Table E.4) and the information collected on the axis of the HED data of the population and the city of Bhopal enables us to highlight *dissonances* between the data axes of each HED.

– Site selected because of the train station, the production of electricity and the presence of sand. – Site 1500 m from the train station, close to the city center and adjacent to a slum. – Very rapid and significant increase in population.

Box E.1. *Data axis of the HED of the city of Bhopal*

– Several thousand people lived in the immediate vicinity of the plant in a slum.

– Many victims died on arrival at the hospital but doctors did not identify the hazardous substance.

– Few victims were compensated.

Box E.2. *HED population data axis*

A first *dissonance* linked to urbanization is highlighted. The city chose the site for its proximity to the station. The site was therefore also close to the city center and a slum. The wind dispersed the toxin towards the nearby inhabited area. This dissonance can also be cross-referenced with the rule identified in the authorities' HED for the period 1957–1980: "It is forbidden to set up a factory discharging toxin products near densely populated areas".

Another *dissonance* appears: the factory doctor indicated that MIC was not a deadly gas. The doctors who treated the victims were unable to identify the toxic (HED of the population) even though the antidote was known.

NOTE.– given the list of deficits and dissonances has been detailed for the different actors chosen for the example, it was not considered useful to list them again using the deficits and dissonances matrices.

E.1.10. *Union Carbide (UC) over time and proposals for action*

The purpose of UC evolved according to the time horizon studied:

– 1957–1980: building the factory in Bhopal;

– 1980–1984: maintaining profitability despite failure to meet objectives;

– Summer 1984: dismantling of the Bhopal plant.

This cindynics approach shows that the strategic choices made from the outset of the study of the Bhopal plant (production capacity of 5,000 t instead of 2,000 t) were one of the decisive factors capable of creating very high vulnerability in the situation. Secondly, wanting to transpose processes linked to American culture, without trying to adapt them to Indian culture, was a very serious mistake. Moreover, from 1983 onwards, no manager had sufficient knowledge of the dangerous nature of the manufacturing process. These few elements show, as one of the participants in the working and reflection group indicated, that the chessboard on which the drama was to unfold was gradually being put in place from the 1970s onwards.

The only way to maintain a high level of plant reliability over the years would have been to assess the impact of the transformation operators on the activity situation studied. This could be done by analyzing the HEDs and then by implementing action plans. A perfect coherence (HED overlay) could have guaranteed the safety of the plant.

As an example, we can propose concrete measures that would have made the HEDs evolve to restore a safer situation:

– 1957–1980: benchmarking on the methods of storing dangerous products, dispersion calculation, associated risks and adapted urban planning;

– 1980–1984: risk assessment and communication approach, use of feedback on accidents onsite, maintenance of safety systems (MIC refrigeration, flare towers, etc., maintaining a preventive and curative maintenance plan);

– 1984: ensuring appropriate communication on the plant's situation, adapting safety rules to changes in production, complying with operating rules (amount of MIC stored).

This cindynic approach was able to integrate all of these parameters, leading to an evaluation of the impacts of the transformation operators and guiding the prevention plans to adapt to the changes. An *a priori* implementation of the approach would also have made it possible to anticipate and adapt the situation to the future transformation operators known within the organization (demographic analysis, calculation of dispersion, storage of the MIC, etc.).

E.2. The Queen Mary II gangway accident

E.2.1. *Context*

Chantiers de l'Atlantique (CAT) was responsible for the construction of the Queen Mary II vessel. The company had previously experienced difficulties in meeting its workload and had implemented a plan to reduce costs while seeking to improve quality and innovation.

The Queen Mary II – in the final phase of construction – had just returned from a second series of sea trials before delivery and was being dry-docked to solve a problem with the propeller.

To continue the finishing work, CAT needed access to the vessel. It asked its subcontractor to install two gangways to serve the front and rear openings of the vessel. The subcontractor was Endel, a company specializing only in the manufacture of scaffolding.

At 2:00 p.m. on Saturday, November 15, 2003, painting work was being carried out on one of the gangways; both workers and visitors could therefore only gain access to the Queen Mary II via one entrance.

Expertise and additions to this expertise demonstrated very clearly that the gangway had collapsed due to physical instability and not due to mechanical failure. The experts pointed to errors in the design of the gangway, as well as errors during its assembly.

E.2.2. *The accident*

A tradition, under certain conditions, allowed the managers and supervisors of Chantiers de l'Atlantique, and those of the subcontracting companies, to show the vessels under construction to relatives, in the evening or at weekends. On Saturday, November 15, 2003, the day of the visit, two groups of visitors accompanied by managers were present to visit the Queen Mary II.

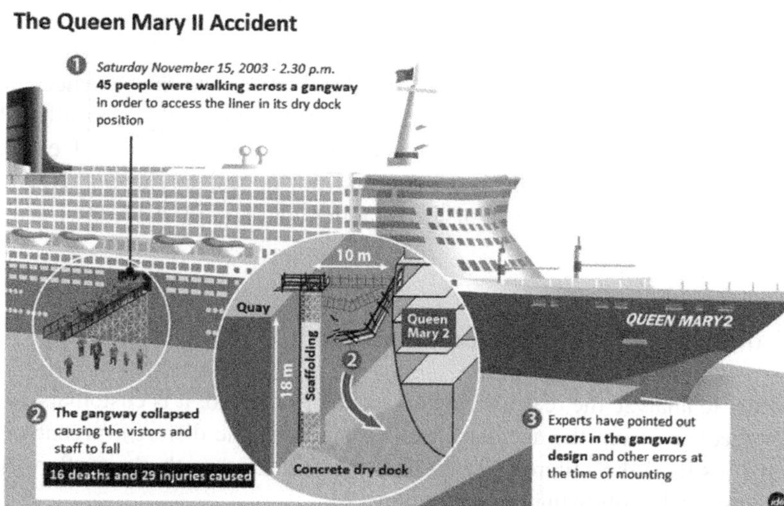

Figure E.1. *Figure showing the gangway accident (source: France Bleu). For a color version of this figure see, www.iste.co.uk/planchette/cindynics.zip*

On that day only one gangway was in service. Around 2:15 p.m., the two groups were on the gangway to access the liner. A third group of employees from the cleaning company responsible for the electrical cabinets, under the direction of a manager from the shipyards, also used this gangway.

Forty-six people were on the only gangway available to access the Queen Mary II. At that moment, a creaking sound occurred, the gangway seemed to collapse in the middle, drop onto the quay side, tip over on the ship side, and fall to the bottom of the hold 18 m below.

E.2.3. *The consequences*

Of the 46 people, 16 died and 30 were injured, 18 of whom were totally incapacitated for more than three months.

The *plan rouge* (emergency response plan) was immediately put in place and the injured evacuated to hospitals in the region.

A judicial inquiry was launched. In the first instance, in February 2008, two executives were released, as were the other six individuals prosecuted in the case. Only Chantiers de l'Atlantique and their subcontractor Endel were convicted as legal entities, each with a fine of 177,500 euros. On appeal, the fine was increased to 225,000 euros.

In 2009, however, the Rennes Court of Appeal confirmed the sentences handed down to the two CAT salaried executives and requested by the Versailles Public Prosecutor's Office. In addition, it convicted four individuals: two CAT employees and two Endel employees. The latter two did not appeal to the Supreme Court, which is what the two executives of the Chantiers did and in 2010, their sentences were overturned.

E.2.4. *Analysis of the causes of the accident*

In order to analyze the reasons why an accident occurred, it is customary to bring together feedback from the accident itself to indentify the different precursors that may have justified the failures. To do this, the fault tree method[2] is often used in order to answer the following three questions:

2 Method developed in the 1970s by the INRS.

– What did it take to make this happen?

– Was this necessary?

– Was this sufficient?

The judicial investigation made it possible to trace the various deficiencies in the design, manufacture and installation of the gangways (see section A9.4). In the case of this accident, the investigation resulted in the establishment of a fault tree (see Appendix 10).

Here, we find the conclusions of the forensic experts regarding "errors in the design of the gangway, added to which are errors at the time of its assembly". The focus was only on the technical aspects since it was only over a short period, from the end of October to Saturday, November 15. Moreover, the conclusion did not take into account cultural or managerial aspects, and even less the temporal aspects that could explain the various failures.

E.2.5. *The cindynics approach to the gangway accident*

On the strength of its experience with the Bhopal study, the working group immediately deemed it useful to conduct documentary research detailing the entire context of this case. To do so, the group relied on the information gathered in [NIC 12]. Within this book, Appendix 3 – entitled "Qualitative Analysis of Several Judgments" – describes the case of "The Queen Mary II Gangway or the Consequences of Design Defects" (see Appendix 9).

Collection of this information made it possible to better establish the activity situation, its context and all the actors concerned.

A sub-group undertook this study following the approach used in the Bhopal case study and presented the following results.

E.2.5.1. *The activity situation*

The system studied: the construction of a prestigious vessel (Queen Mary II) by the shipyard "Les Chantiers de l'Atlantique" (CAT) located in St-Nazaire, France, within the framework of respect for the quality, the deadlines and the potential costs of the order.

The context: the St-Nazaire site has been managed since 2006 by the Scandinavian group Axe Yards following a stream of sales, from Compagnie Générale Transatlantique to Axe Yards, via Alstom. Competition in the shipbuilding sector is fierce, which led CAT to decide to implement two successive cost-cutting plans which began in 1997. This decision led to a change in their organization. CAT chose to refocus on its core business (naval architecture, boiler making and assembly) by subcontracting other activities that were outside core business. This choice led to coordination between CAT and subcontractors.

This new set-up meant that CAT needed to formulate its need, receive a quote, evaluate it, place an order, and then receive the product.

The chosen subcontractor needed to study the request, draw up its estimate, then on receipt of the order, elicit what was needed for manufacture, assembly and inspection before final acceptance of the product by the customer.

It should be noted that the need at the origin of the accident consisted of the installation of two gangways in order to transfer packages from the dock to the ship, using a machine weighing 1.2 tons, as well as workers and visitors of the ship under construction. Within the framework of a set of specifications, this involved having these two gangways built and installed by a subcontractor, and then used.

The spatial and geographical horizon: delimited to the cruise ship construction sector in Saint Nazaire by the company CAT, to the quay, to the ship in dry dock and to the gangways.

The time and chronological horizon which cover:

– an initial period between the end of 1997, when CAT implemented its new set-up, and September 2003, when the first sea trials of the Queen Mary II took place;

– a second period running from October 2003 to the end of the trial, comprising six phases from the definition of the need for gangways to the accident, followed by the trial.

The value of having chosen to explore the horizon since 1997 was justified by the change in the organization of the company with the increase in subcontracting, creating a multitude of new interfaces conducive to ambiguities and misunderstandings. It also seemed useful to extend the horizon until the end of the trial, as it allowed the company to benefit from the information drawn from the "interrogations". This possibility reflected the need to conduct interviews with the actors in order to fully understand, rather than just interpret, the information gathered.

Networks of actors: as with Bhopal, the choice was made according to the temporal, spatial and contextual criteria mentioned above.

The *actors* identified for the first period were as follows:

– for CAT: management, those responsible for work on board, design offices, coordinators and safety managers;

– for ENDEL (subcontractor): the branch manager, the project managers, the design office and site managers;

– for ISP (subcontractor): those responsible for site safety.

To which should be added for the second period:

– users: cleaning staff and visitors.

Of all the information gathered (see Appendix 9), the last two types of actors cannot be considered relevant, because even though they were involved in the event, it must be noted that they were not responsible for the fatal outcome.

The following *relevant actors* were therefore retained:

– For CAT:

- management;
- work coordinator;
- head of on board works;
- on board works manager;
- head of logistics methods.

– For ENDEL:

- head of agency;
- design office manager;
- site foreman coordinator.

During the first period, CAT placed an order for external scaffolding for the upper part of the vessel (tender batch 1).

For the second period, CAT placed an order for additional work including gangways. As this represented a significant part of the vessel's finishing activities, it is possible to compare the baseline and observed activities.

Key activities: the definition of the need had to be developed by the project owner and, more specifically, by the methods department which then played the role of client delegate. The latter needed to specify:

– the function of the gangways, either limited to the passage of handling equipment and packages, or also allowing access on board for intervening personnel and visitors, as the regulations were different;

– the distances from the platform, the dimensions (length, width), the loads to be supported, the means of anchoring the gangways;

– delivery times and places.

The purchasing department selected suppliers potentially able to meet this demand according to their skills, then upon receipt of the offers, and with the help of the project owner's delegate, analyzed the proposals in order to make a choice.

When placing the order, the supplier drew up calculation notes, determined the general structure of the product and carried out the procurement necessary to produce the product in accordance with the request.

The assembly needed to be strictly in accordance with the proposal and at the end of the assembly, the client delegate accepted the product by verifying its conformity to the request.

Observed activities: according to the elements collected (see Appendix 9), practically none of the actions cited as key activities were followed up. The ordering process was almost non-existent as CAT managers offloaded almost all of their tasks to ENDEL. The construction plans were imprecise, which led to inverting the locations of the gangways and changing the width requested.

Due to the lack of precision in roles, missions and interface management, the position of CAT managers led to captive subcontracting which induced routine behavior, eliminated any critical thinking and produced non-compliance leading to a serious accident.

The urgency to complete the finishing touches on the boat for its scheduled delivery on December 22, 2003 may also explain these deviations.

The table below summarizes the elements between the baseline situation and the observed situation.

	Baseline situation	Observed situation
Purpose of the study	Analyze the reasons for the collapse of the gangway	
Context	Reorganization of CAT, subcontracting. Ship in dry dock, tolerance of industrial tourism, distinction of the use of gangways (equipment, people).	More than 1000 visitors expected – mixed use of footbridges. Co-activity (maintenance and visit). Only one gangway available on the day of the visit.
Spatial perimeter	Shipyards of St. Nazaire quay, two gangways, ship in dry dock	
Time horizon	**Two periods, six stages:** reorganization of CAT and request for gangways. **Six phases**: definition of the need, order the gangways, design, manufacture, assembly and acceptance, accident and trial.	No precise definition of missions and relations between CAT and subcontractors. No rigor in the processes, from the order to the reception of the gangways.
Activities carried out and recent developments	Construction of a prestigious vessel with respect to quality, costs and deadlines. Order to ENDEL in 2002 for interior and exterior scaffolding for the boat. Construction of two gangways through subcontractors, according to specifications dedicated to the intended use (equipment and personnel).	Inaccurate specifications, work without a plan, acceptance of an estimate that did not comply with the request, lack of calculation and acceptance notes. Inversion of the gangway and modification of one of them (*in extremis*), lack of skills in the realization of the gangway (accepted by the client) and its installation.
Actors	Period 1: CAT (project coordinator, head of methods, management), ENDEL, ISP (in charge of access controls). Period 2: cleaning staff and visitors.	Periods 1 and 2: the same.

Table E.5. *Comparison between baseline and observed situations*

Qualification of the risk sources: as in the case of the Bhopal disaster, this will be established based on the overall level of the organization as well as on the five axes of the **hyperspaces of danger** associated with the actor networks.

E.2.5.2. *Systemic deficits (CSDs)*

A few CSDs can be noted (see Appendix 5):

CSD 1 – *infallibility culture*: not apparent.

CSD 2 – *culture of oversimplification*: underestimation of demand, ENDEL specialized only in industrial gangways; use of catalogs to design gangways for various uses, no calculation notes, no plans to rebuild the gangway; reception of gangway following a quick glance.

CSD 3 – *communication culture*: lack of documentation and contractual communication between all the actors involved; an operator removed two forms of bracing without referring to the assembly manager.

CSD 4 – *navel-gazing culture*: not apparent.

CSD 5 – *priority given to production over safety aspects*: the vessel was at the end of construction, it needed to be delivered quickly; all actions were carried out in a hurried manner and took priority over safety.

CSD 6 – *dilution of responsibilities*: CAT transferred to its role as project owner to ENDEL; the design of the gangways was entrusted to a manager who was not qualified; nobody felt responsible for their own mission.

CSD 7 – *lack of feedback*: some activities were carried out without plans and recollection.

CSD 8 – absence *of procedure with a cindynic dimension*: absence of workstation sheets, interface relations between CAT and its subcontractor were not defined, absence of written procedures; no respect for the passage process on the gangway and control of the order; absence of use of working methods.

CSD 9 – *lack or insufficient knowledge in risk management*: the order specified a load of 250 N/m^2, the subcontractor proposed a structure for 150 N/m^2 and this solution was accepted by the client!

CSD 10 – *lack of preparedness for crisis situations*: no elements in the text.

E.2.5.2. *Hyperspace of Danger (HED)*

Among the actors considered relevant, the following will be chosen as examples:

– For CAT: management and the coordinator of works.

– For ENDEL: the head of the agency and the construction site manager.

The hyperspaces will be presented in the form of tables, showing the different possibilities of schematization of the HEDs.

The “Expected” level corresponds to what was desired hypothetically (see Chapter 1, section 1.3) and the “Observed” level corresponds to the collected elements (see Appendix 9).

		Purposes	**Values**	**Regulations and standards**	**Facts and data**	**Models**
CAT – Management	Expected	– Deliver a vessel on time, on budget, high quality and safety. – Objectives to further reduce costs and improve quality and innovation.	– Tradition of showing ships at the end of construction.	– General rules of a company. – Assignment of roles and delegations, particularly in the context of outsourcing calls for tenders.	– Placing an order with approved and competent suppliers.	– Management of interfaces between MOA and internal subcontracting. – Shipbuilding benchmark.
	Observed	– Priority of costs and deadlines over production and safety.	– Respected tradition.	– Ignorance of the attributions of managers. – Non-compliance with standards.	– Order of gangways at Endel, (scaffolding). – Estimate that the accident was the result of a design and construction defect. – Imprecise roles and tasks, both in terms of CAT and relations with ENDEL.	– Lack of knowledge of interface management concept.

Table E.5. *Actor: CAT management*

CAT – works coordinator		Goals	Values	Regulations & Standards	Facts & Data	Models
	Expected	- To ensure the construction and ancillary services of the ship while respecting safety, quality, deadlines and costs.	- Respect for cost, deadlines, safety and performance. - Exemplary nature, rigor of realization.	- Safety referent (on-site delegation). - Ensure the application of the site safety regulations and labor code.	- Authority over supervisors, supervising CAT employees and subcontractors. - Financial powers. - Health and safety delegation.	- Labor code and framework for establishments open to the public. – Health and safety regulations. – Organization of site management with respect to technical and regulatory constraints.
	Observed	- No subscription to its commitments as purchaser. - Did not care that only the C292 gangway was available. - General laxity. - Priority was given to meeting deadlines.	- No verification of the quality of the assembly of the gangways. - Delegated safety. - Trust put in teams without any form of control. - Did not feel concerned about the causes of the accident.	- Ambiguity on the gangway names, between gangway and scaffolding. - Unclear on its scope of responsibility in terms of safety.	- Allowed installation without bracing. - Spacing of visits not organized. - Access coordination neglected. - Absence of reception tickets.	- Non-compliance with regulations for establishments open to the public. - No visible model of site management then followed.

Table E.6. *Actor: CAT works coordinator*

ENDEL – head of agency		Purposes	Values	Rules and standards	Facts and data	Models
	Expected	- Obtain contracts to manufacture industrial gangways (scaffolding). - To provide satisfactory answers to their client's questions.	- Ensure that the urgency of the order does not compromise the quality of the service. - Expertise and rigor.	- Respect of the rules for the construction of gangways.	- Verify that all technical and financial aspects have been taken into account before validating the order.	- Contracting between different stages. - Technical model and knowledge of material strength calculation.
	Observed	- Financially oriented. - Failure to take into account the specifications of the application.	- Validated the estimate in the absence of a business engineer. - Priority of financial aspects over technical needs.	- Did not ensure compliance with the rules of the trade and the contractual provisions binding the company to CAT.	- Control of financial and administrative aspects. - Validated the estimate without taking note of the nature of the need, the calculations and the plans.	- Non-compliance with contracting stages - Sizing of gangways taken from scaffolding catalogs.

Table E.7. *Actor: head of the ENDEL agency*

ENDEL – site manager		Purposes	Values	Regulations and standards	Facts and data	Models
	Expected	- To provide a product in conformity with the order by respecting the parameters of cost, delay, quality, etc.	- Compliance and rules of the art. - Quality of the service.	- Compliance with technical standards and assembly of gangways.	- Capacity, resistance, length and width of the ordered gangways. - Correct location, stability and evenness.	- Professional expertise in sizing and assembly of gangways.
	Observed	- Improper delivery and assembly. - Respect of the deadline.	- Do-it-yourself, amateurism, lack of skills.	- Superimposition of the sheet metal resulting in additional weight. - No bracing (to allow assembly).	- Reverse assembly of batches 413 and 418. - Not surprised by the absence of quotation. - Incorrect location, stability and evenness. - No written procedures.	- Improvisation instead of expertise.

Table E.8. *Actor: ENDEL site manager*

E.2.5.3. *The identification of deficits and dissonances*

When studying these tables, the deficits and dissonances appear numerous.

There are many *deficits*:

– *Regarding the CAT management*:

- *gaps*, particularly the case after a reorganization to refocus on the core business of the site, they ignored the real responsibilities of their managers and the management of interfaces between their units and sub-contractors who worked on the same site as the main activity;

- *degeneration*, no mention of prioritization between a focus on cost reduction and a focus on quality and safety. The priority of cost reduction was chosen;

- *disjunction*, between the goal of improving quality and the lack of a model for interface management.

– *Regarding the CAT works coordinator*:

- *gaps*, due to a lack of time or knowledge, he transferred commitments from the client to the subcontractor. In addition, there were shortcomings in the mission to authorize visits. The coordinator did not manage the flow of visitors, with the aggravating element being that there was only one form of access to the ship available. In addition, he ignored the existing regulations for establishments open to the public;

- *disjunction*, between its mission to ensure safety and the lack of control taken during the installation of a gangway without bracing;

- *degeneration*, by being imprecise in the names plank, gangway and scaffolding.

– *Regarding the head of the ENDEL agency*:

- *gaps*, because he only focused the validation of his company's offer on the financial and administrative aspects, without being aware of the nature of the need and the calculations;

- *disjunction*, between purposes and rules, because he did not ensure the respect of the contractual rules that bound his company to CAT, as well as between purposes and values, since he validated the file in the absence of his business engineer.

– *Regarding the ENDEL site manager*:

- There were many gaps; they were numerous on all qualifiers, delivery, assembly, rules and data, as well as models. These were visible on the representation of the hyperspace relative to the site manager.

As for the *dissonances*, there were two types:

– the first, which was probably the cause of a misunderstanding, was due to the ambiguity of the terms gangways, planks and scaffolding. The order placed mentioned gangways, but its description only referred to the loads to be transported, whereas there was a tradition of showing ships at the end of construction. In addition, this gangway was to be used to transit subcontracted personnel working inside the ship;

– the second relates to the representations of each of the parties after the accident. For CAT's officials, the accident resulted from a defect in the design and construction of the gangway, while for ENDEL's officials, the accident was due to the use of the gangway under conditions that were neither foreseen nor foreseeable.

This census of deficits and dissonances allows us to draw up matrices of deficits and dissonances. However, since only a few actors have been treated as examples, it was not considered necessary to transfer the preceding elements to matrices.

It is nevertheless possible to deduce an action plan from the identification of deficits and dissonances.

Reduction of deficits:

– *For CAT management*:

- take over the organizational procedures (missions, delegations) according to the core business and subcontracting activities;

- prioritize between cost reduction and quality-safety objectives;

- manage the interfaces between the project owner and subcontractors better;

- clarify the role of the health and safety manager;

- manage the adequacy of skills/functions;

- raise awareness of safety culture.

– *For the CAT works coordinator*:

- clarify the vocabulary between gangways, planks and scaffolding;

- receive training in safety according to the delegations of missions.

– *For the head of the ENDEL agency*:

- receive training to strengthen their technical skills;

- take a training course in contract management.

– *For the ENDEL site manager*:

- receive training in compliance with organizational procedures.

Reduction of dissonances between CAT and ENDEL:

– organize site meetings according to the different stages.

The accident on the Queen Mary's gangway is a good illustration of the impact of silent transformations on the same site, generating a propensity towards disaster, a propensity amplified by the lack of efficient management of the interfaces between the contracting authority and subcontractors.

E.3. The Deepwater Horizon drilling rig accident

E.3.1. *Context*

Oil exploitation in the Gulf of Mexico is subject to very strong competition between several operators. In addition, the offshore exploitation zone is located in the deep waters near Louisiana, at the heart of ecosystems rich in biodiversity [MAG 10a] (numerous aquatic species, coral beds, etc.).

In this context and in order to improve the performance of the field teams, BP has adopted management methods that leave more initiative to the field teams. In March 2008, BP acquired the drilling permits for the Macondo well. During 2009, BP leased a platform (Marianas) from Transocean to drill the Macondo well.

Drilling began with this platform, which was damaged by the cyclone (Ida) and then evacuated in November 2009. At the end of January 2010, Transocean then put in place a new dynamically positioned platform "Deepwater Horizon".

This platform had to complete the drilling and secure the well (sealing by cementing) by the end of April 2010 in order to be available for another drilling. Another platform was then to be put in place to finish equipping the well and start production.

However, due to the complexity of the layers to be drilled, the project was more than a month behind schedule.

E.3.2. *The accident*

On April 18, 2010, the cementing operation of the well was completed and considered satisfactory. The cement was supplied by the Halliburton company. However, the BP project team did not require, as foreseen by the procedures in force, the three main complementary tests to verify the quality of the cement. Although some of the leakage tests carried out were subject to divergent interpretations that did not allow the conclusion to be drawn that the well was watertight, the drilling team believed that the installation of the cement plug would be sufficient to check the watertightness of the well. Unfortunately, this was not the case.

On April 20, pressurized hydrocarbons entered the bottom of the well. Following the failure of the Blow Out Preventer (BOP), these hydrocarbons rose into the well casing, unbeknownst to of the rig crew.

Mud and hydrocarbons ended up gushing out onto the platform's deck. The gas spread, ignited and exploded. The platform could no longer be controlled by the emergency devices whose remote controls had become inoperative.

These conditions generated an atmosphere of panic and the decision was made to evacuate the burning platform. Unfortunately, the oil leaks continued to foul the waters of the Gulf of Mexico and were not brought under control until early August.

E.3.3. *The consequences*

Eleven people were reported missing. The submarine eruption, which continued until August, was only stopped thanks to the action of a remote-controlled submarine intervening on the leak. This event resulted in the most serious marine pollution in the Gulf of Texas. The oil slicks polluted the ecosystems. "But the deepest and most serious impacts probably concerned deep marine ecosystems, by definition not very visible [...] and where cold temperatures and lack of light did not promote biodegradation" [MAG 10a].

Figure E.2. *Image from the Final Report on the Investigation of the Macondo Well Blowout. For a color version of this figure see, www.iste.co.uk/planchette/cindynics.zip*

E.3.4. *The cindynics approach*

A sub-group was responsible for the cindynic study of the platform accident, following the same approaches adopted for the two previous examples.

E.3.4.1. *The activity situation*

The cindynics approach to an accident implies that one follows its processes (see Chapter 1, section 1.4) by first gathering information, both on the events that preceded the accident and on the development of the accident situation. This study was undertaken through reports, books and a film. The information gathered (see Appendix 11) only allowed a partial analysis of the facts and behaviors of the different actors involved. Due to not having an analyst linked to the drilling profession, some information could not be deciphered to allow a complete study of the activity situation, and compare what was expected from an actor (specifications, rules and practices of the profession) compared to what could be observed before and during the development of the accidental situation.

This remark highlights the need to conduct interviews or in-depth analyses of documentation in sufficient quantity and in the company of experts from the various professions. This is the only way to carry out a global cindynics study in order to discern what, in the behaviors of each actor and between actors, would have highlighted possible deficits and dissonances that could be risk sources.

Since it was not possible to steer the cindynics approach at this level, the approach was therefore limited, on the one hand, to looking at the global level of the organization of the platform's activity in order to evaluate the systemic deficits and, on the other hand, to the study of a single actor's behavior (the client's representative on the platform).

Even though this analysis was incomplete, it nevertheless made it possible to identify certain risk sources and propose some preventive measures to reduce them.

E.3.4.2. *Systemic deficits*

The analysis of systemic deficits was made possible by integrating the various observations noted by the authors of the various information sources consulted (see Appendix 11).

Cultural	**Infallibility:** seven years without accidents, price of safety, blow-out denial.
	Oversimplification: relying on an analysis of causes without an overall analysis of the system; ignoring the abnormal course of the well depression test.
	Non-communication: Transocean/BP deficit; within the Transocean team; Halliburton/Transocean; between managers and staff.
	Navel-gazing and lack of vigilance: ignoring or pretending to ignore accidents on other platforms.
Organizational	**Productivism before safety:** improving basic productivity, impasses regarding leak testing operations, a need to catch up.
	Poor definition of responsibilities: especially in crisis management, discussions between stakeholders and operators without a decision-making process; poor management of priorities; poor control of the management team.
Managerial	**Lack of experience feedback:** no consideration of previous accidents, even recent ones, in particular the one that occurred one month before the explosion.
	Lack of safety training: weakness of the BP Group's centralized safety policy.
	Lack of preparation for crisis situations: barriers and defense systems did not work; lifeboats remained attached to the structure; no energy back-up.
	Lack of awareness of risk management: weak safety culture at BP; analyses made at the level of individual operations and not at the system level.

Table E.9. *Systemic deficits*

E.3.4.3. *Actors involved*

The reading of the various documents made it possible to identify several actors who played a role before or during the operations that led to the accident:

– The control authority: **Minerals Management Service (MMS)**, a department of the Ministry of the Interior responsible for granting permits and controlling operations.

– The project owner: **BP**, responsible for the workflow and contractualization of the various operators. BP personnel were present on the platform.

– The prime contractor for the well drilling operations: **Transocean** was contracted by BP to provide the platform and the personnel responsible for the drilling and maintenance operations, and had BP's delegated authority for the supervision and control of all of the work.

The operations' players were as follows:

– **Halliburton:** in charge of all work related to the cementing of the well to ensure its tightness and prevent hydrocarbons from migrating to the surface.

– **Swaco:** in charge of supplying the drilling mud.

– **Sperry Sun:** Halliburton subsidiary in charge of drilling mud and pressure control.

– **Schlumberger:** in charge of all well logging operations and monitoring of the physical characteristics of the layers penetrated.

– **Oceanring:** in charge of the inspection of the submerged material: wellhead, BOP and its surroundings using a remote controlled robot.

The players regarding equipment were as follows:

– **Weatherford:** in charge of supplying the equipment (casings) for the well based on the specifications provided by BP.

– **Dril-Quip:** in charge of supplying the wellhead, its installation and maintenance.

After making an inventory of the actors involved, the sub-working group selected those it considered most relevant by their involvement and responsibility in the events leading up to the accident.

In these choices, those of the project owner BP and the project manager Transocean were considered relevant. Based on the information collected, we will then focus on a cindynics approach for the BP client's behavior.

E.3.4.4. *The hyperspaces of danger (HED) realized*

The "expected" and "observed" behaviors of the representative on the client's BP platform are presented below.

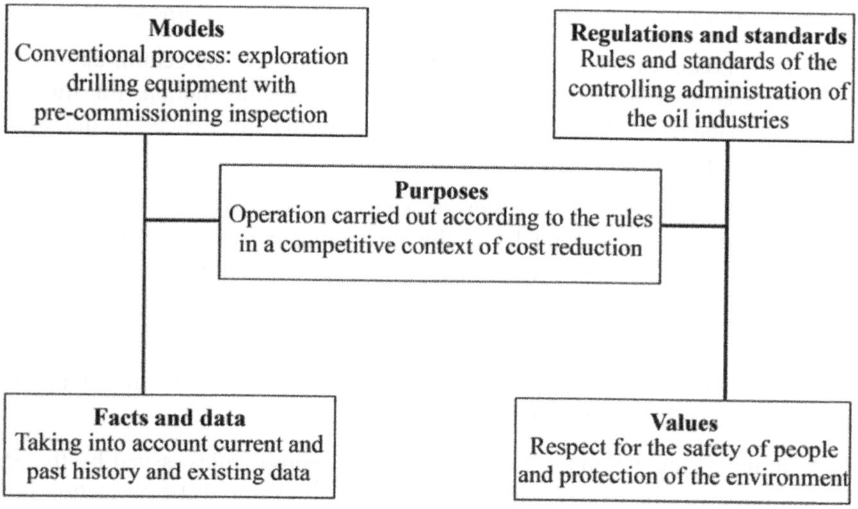

Table E.10. *Actor: BP – expected behavior*

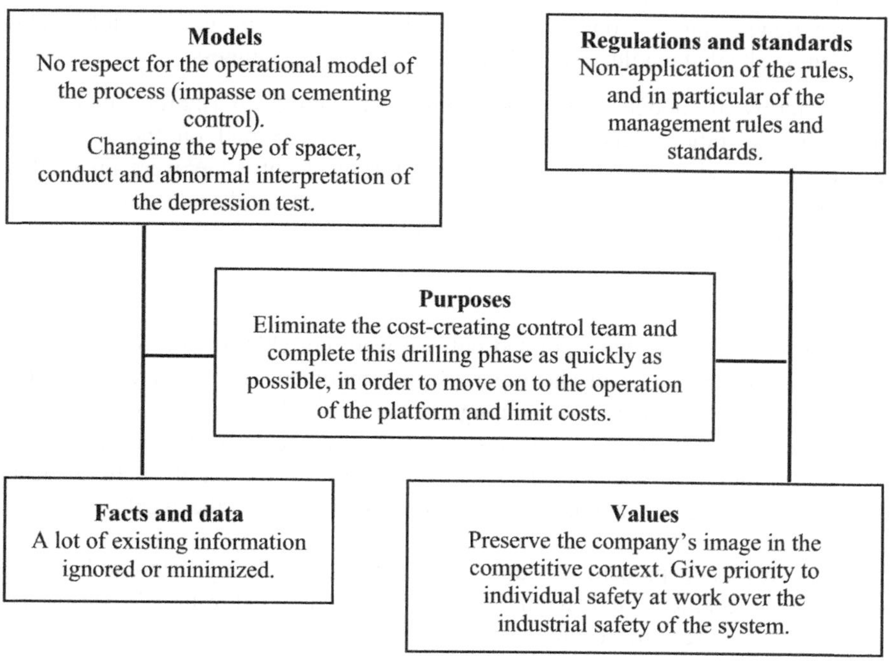

Table E.11. *Actor: BP – observed behavior*

E.3.4.5. *Statement of the actor's deficits: representative of the client*

The comparison of what was noted in the behavior of this central actor, when faced with what should have constituted the basics of the instructions transmitted to the project manager (TRANSOCEAN), showed strong differences in the various "cindynic" components:

– In terms of purposes: the requirements of cost reduction and respect of deadlines moved to the second level, the final objective of completing the work stage by applying the elementary rules of the profession in order to obtain a watertight and safe well to carry out the next stage of exploitation.

– In terms of values: the overall safety of the platform (and not only those relating to safety at work) took a back seat in order to preserve the image of a company that respects its costs and deadlines.

– In terms of regulations and standards: the numerous rules issued by the State Administration for the overall safety of drilling platforms and the protection of the environment had been violated (in particular, those relating to managing the watertightness of the well cementing).

– In terms of facts and data: it seems that the operator did not take into account the experience gained from other well drilling operations, nor the information gathered on site that would cast doubt on the well's watertightness.

– In terms of models: such operations were subject to rigorous models for conducting operations, and observations made and reported on the conduct of operations have shown that impasses and unfortunate initiatives were undertaken, deviating from the rules of the trade. There was no physical understanding of the phenomena, only one competent person at the operations level.

E.3.4.6. *Some considerations from the global analysis of the actors' behavior*

Even though, as we have pointed out, we were not able to carry out a detailed analysis of the behavior of the other players involved, the information sources consulted show that many shortcomings existed, well before the accident, in the transmission of information and the coordination of the teams of the various players involved. Centralized and well-informed management is certainly a fundamental element for the safe conduct of such operations, particularly in the event of a crisis.

This cindynics approach, although incomplete, has made it possible to highlight, beyond the technical and natural causes, several organizational causes and coordination failures, the accumulation of which is at the origin of the development of the accident and its disastrous consequences in terms of loss of human life, environmental degradation and economic waste.

The priority given to economic and time considerations, as well as the focus on the safety policy of the contracting authority, shows a global governance flaw, the effects of which were particularly catastrophic.

E.4. Covid-19 and the problem of home-made cloth masks

E.4.1. *The general context*

E.4.1.1. *The facts*

This latest pandemic, known as Covid-19, began with numerous cases of contamination detected in China's Hubei province of Wuhan as early as December 2019, before spreading to the rest of the world from the beginning of 2020. It should be noted that, according to some sources, the virus appeared in Colmar, France in mid-November 2019.

The appearance of Covid-19 was seen as a new type of risk, whereas many other viral risks (Spanish flu, Hong Kong flu, SARS, H1N1, Ebola) have already highlighted the emergence of epidemic risks.

Indeed, a century ago, the Spanish flu (1918–1919) killed some 50 million people worldwide. The experience was that this type of respiratory infection was transmitted by:

– droplets expelled when people cough, sneeze, or even breathe;

– contact with a contaminated surface.

As soon as this pandemic began, the wearing of masks was imposed on anyone who found themselves in public places in San Francisco, under penalty of a fine, or even imprisonment [BEA 07]. The report cited mentions that "the wearing of masks is part of the hygienic practices of modern medicine!".

In Europe, other epidemics have occurred. And the French article "*Ce que les pandémies du passé peuvent nous apprendre sur les dangers du COVID-19*" (What past pandemics can teach us about the dangers of Covid-19)[3] shows that preventive

3 See: https://www.franceinter.fr/histoire/ce-que-les-pandemies-du-passe-peuvent-nous-apprendre-sur-les-dangers-du-covid-19.

measures have always been taken at the time of crises in the past, but unfortunately they are quickly forgotten. In particular, Patrice Debré (a French medical immunologist and academic) reminds us that: "It is easy to forget and it is one of the duties of scientists to try to continue to explain to people that **the emergence of an epidemic is always unpredictable and that we must constantly prepare for it** [...] in non-epidemic times, learn about and prevent them". As a reminder, we will recall the 2 billion masks stored in France in 2009 that have since "disappeared"!

This new Covid-19 virus, whose less virulent form would have taken over before spreading, has a special characteristic. It causes a disease whose symptoms do not immediately appear. This phenomenon has brought a novel risk scenario to light, based on an original and previously unobserved mode of contamination, namely, the transmission of the virus by healthy carriers.

Virtually all countries were therefore surprised by its scale and speed of spread.

A few countries have responded quickly based on feedback from previous peak epidemics and, as a result, have been able to better contain the pandemic. But others were overwhelmed by the rapid spread of the virus. Some countries downplayed the danger by deliberately choosing not to implement barrier prevention measures. Others even chose to oppose them, but were forced to do so after the same leaders who recommended it became infected!

The spread of the virus then became global, and depending on the strategies used, the health and economic outcomes have been quite different. For example, thanks to its remarkable crisis management, Taiwan has one of the best human balance sheets in the world and has been able to maintain its economic activity despite the pandemic.

Moreover, in some countries, the bond of trust between the population and the state is now under threat. This does not predispose people to accept the barrier measures taken to fight the spread of the coronavirus: keeping a safe distance, washing their hands regularly with soap, not touching their face and wearing a medical grade mask [MAY 20]. This loss of trust may even be accompanied by uncontrolled fears.

E.4.1.2. *General prevention measures against the pandemic*

First, we will develop a statement of the actions currently being taken. What do we find?

– the danger has been – at least in some Western countries – poorly appreciated, regarded as unknown or unimportant because of the predominance of a sense of infallibility that increases the inability to contemplate the worst. As a result, the stage of assessing the risk of virus transmission by droplets projected by sneezing or coughing has not (or only to a limited extent) been undertaken. However, feedback from previous epidemics confirmed the transmission of the virus by these droplets.

Dupuy reminds us that "the terrible thing about a disaster is that not only do we not believe that it will happen, even though we have every reason to know that it will happen, but that once it has happened it appears to be part of the normal order of things. Its very reality makes it commonplace":

– The Covid-19 "epidemic" began in China in December 2019, but the WHO did not sound the "pandemic" alarm until March 11, 2020!

– The WHO and government doctrine regarding the recommendation of masks to be worn by the public evolved during the coronavirus outbreak. In the early days, WHO claimed that masks were unnecessary. France followed suit by claiming, during March 2020, that masks were unnecessary for the public. They then changed this position on May 11, making it mandatory on public transport, in certain working circumstances (see the OPBTP guide from April 15) and recommending it in areas where several people may have close contact. This use complemented the barrier gestures and rules of physical distancing, especially for healthy or asymptomatic people.

– In some companies, the internal prevention system was not organized in time.

– The labor code required companies to do so, and the preventive function was very effective in some organizations.

– The study stage on the vulnerability of human and economic issues was insufficiently completed, which delayed the implementation of *real prevention measures*.

These various points led, in the midst of a crisis and a shortage of masks and respirators, to urgent *protective measures* being taken in the health system (mobilization of hospital beds, a race to obtain masks and respirators, etc.), and at the social level (physical distancing measures, lockdowns, masks, etc.), as well as at the economic level (company closures, short working hours, working from home, etc.).

Without business continuity plans, some states and companies believed they could manage the crisis by making decisions in a climate of near panic.

According to the essentially technical (lack of respirators in Lombardy) or material limitations (masks, protective clothing), two options were offered to the leaders:

– testing and isolating patients and subjects that they had been in contact with: this was the case in Taiwan, Korea, and Germany;

– immobilizing the entire population (by reproducing the strategy that China had deployed for only 6% of its population: that of Hubei).

E.4.1.3. *Specific measures for the wearing of masks*

E.4.1.3.1. Regulations and directives in France

For nursing staff, the wearing of sanitary masks (FFP2 and surgical masks) is mandatory.

New categories of masks reserved for non-health uses were defined by the interministerial note of March 29, 2020.

These are two new categories of alternative fabric masks that can be reused. Category 1 is used by professionals in contact with the public, while Category 2 is intended for collective use, to protect an entire group wearing these masks. This interministerial note specifies as technical specifications the efficiency parameters:

– filtration of the 3 μm particles emitted (the filtration rate is the proportion of aerosol that has not been retained by the filter);

– breathability at a level that allows it to be worn for up to four hours, as the ability to breathe well decreases as the mask is humidified.

The achievement of the efficiency levels for masks for collective use: filtration > to 70% and air permeability (in $Lm^{-2}s^{-1}$) > to 96 for a depression of 100 Pa – must be verified by tests according to various test protocols. These are defined either by the European standard NF EN 149, by notified bodies, by AFNOR (leader of the French standardization system), or by the protocol described in the document of the *Direction Générale de l'Armement* (DGA).

AFNOR has made standard 14683 freely available on its website and published the S76-001 specification at the end of March, which is a guide to the minimum requirements, test methods, manufacture and use for the mass production and hand-made manufacture of these "barrier masks".

For companies wishing to market masks to the general public, obligations are issued by the *Direction Générale des Entreprises* (DGE). They consist of "having tests carried out by a competent third party demonstrating the performance of their masks in relation to state specifications" so that they can affix an identification logo.

Other recommendations have been issued to ensure that masks are worn for no longer than four hours.

E.4.1.3.2. Uses

FFP masks protect the wearer from inhaling virus-laden droplets and also prevent the droplets from being projected.

Surgical masks are an anti-projection type of mask, that is, **they do not protect the wearer, but those around them.**

The model for the use of barrier masks in a pandemic situation has so far been poorly developed, with a tendency to deny the contribution of barrier masks in preventing the spread of any infectious disease: "Cloth (e.g. cotton or gauze) masks are not recommended under any circumstance"[4]. Therefore, they cannot be used.

E.4.2. *The cindynic study of the problem of home-made cloth masks*

E.4.2.1. *Context*

Taking into account the anguish felt by the French population, which – even though its proximity to the Covid-19 virus can be considered less frequent than for healthcare personnel or professionals in contact with the public – the interministerial note from the end of March 2020 specifies new categories of masks for non-sanitary use.

This specification is a reminder that the use of any alternative mask is not a substitute for government guidelines and advice outlined as follows:

– respect boundaries: social distancing, lockdown, application of general hygiene measures, especially hand washing as often as possible;

– be attentive to the duration of effectiveness of this type of mask, which should not exceed four hours;

4 World Health Organization, January 29, 2020: https://apps.who.int/iris/bitstream/handle/10665/330987/WHO-nCov-IPC_Masks-2020.1-eng.pdf?sequence=1&isAllowed=y.

– use masks rationally: in open spaces with a lack of proximity, it is not useful to wear a mask (e.g. wearing an anti-projection mask when you are alone leads to a waste of resources);

– carry out an effective wash at 60° with a detergent, because the danger of contamination, either from the tumble dryer or from a person with secondary contamination, will remain.

Following this interministerial note, a flurry of contradictory recommendations then appeared for the manufacture of masks made from different materials. A multitude of tutorials (on the Internet and by various television channels, press articles, etc.) indicating how to make masks at home appeared. In one video, Prof. Daniel Garin, an expert in biological and infectious risks, learned how to make a mask with a paper towel and two elastics. Another expert explained that it is better to use "good filter cloth" without giving any definition of criteria for a "good filter cloth".

Moreover, various articles state "that a screen in front of the nose and mouth is better than nothing"[5]. Indeed, in 2013, Anna Davies from the University of Cambridge and her colleagues proposed this formula[6]. However, the results showed that even though "these two masks (surgical and alternative) considerably reduced the number of micro-organisms expelled by volunteers, the surgical mask was three times more effective in blocking transmission than the home-made mask". They concluded by suggesting "that a homemade mask should only be considered as a last resort to prevent droplet transmission from infected people, but that it would be better than no protection".

The first journalistic studies by national media (Le Monde on April 6, 2020; France Info on April 7, 2020) were beginning to mention the difficulties in evaluating the effectiveness of masks made from unapproved fabrics.

And yet, masks were beginning to be produced by some textile companies and other companies. The vast majority of industrial productions refer to the AFNOR specification and to the data made available by the DGA on the breathability and filtration capacity of a range of fabrics.

---

5 See: https://www.dna.fr/edition-strasbourg/2020/03/24/une-allergologue-de-strasbourg-un-ecran-devant-le-nez-et-la-bouche-c-est-mieux-que-rien.

8 See: https://www.cambridge.org/core/journals/disaster-medicine-and-public-health-preparedness/article/testing-the-efficacy-of-homemade-masks-would-they-protect-in-an-influenza-pandemic/0921A05A69A9419C862FA2F35F819D55/core-reader#.

However, the reliability of the products made at home has never been tested. Wearing this type of mask can generate a feeling of great security due to the fact that "you have made it yourself". But like any technical means, the alternative mask can be responsible for the worst, as well as the best, depending on how it is made.

Faced with this imbroglio of advice and recommendations, the population found itself perplexed and lost confidence in its leaders and experts.

However, since wearing masks became mandatory in certain circumstances, even difficult ones, a choice needed to be made that would be all the more imperative for families with low resources, who may prefer reusable masks that can be washed. For this solution, care advice has unfortunately not been the subject of an accessible scientific verification.

To reduce all of these uncertainties, **it therefore appeared crucial to implement a process to validate masks worn by the public.**

It was then that the *Société Française de Médecine de Catastrophe* (SFMC – a non-profit association) approached the *Institut pour la Maîtrise des Risques* (IMdR – also a non-profit association) with the aim of conducting a project of specifications for the manufacture of these types of reusable masks.

E.4.2.2. *Introductions to the project*

Before starting this project, whose aim was to improve the performance of masks made at home, **the IMdR perceived the need to conduct a cindynics approach in a pre-accident context.**

Indeed, in this unprecedented period of the Covid-19 pandemic, where the emotional burden is becoming more significant than scientific knowledge, cindynics facilitates a life-saving distance "by questioning stakeholders on the importance of sharing the same **goals, values, data, rules and models** in a situation of uncertainty and disorganization" (Dr Jan-Cédric Hansen).

By placing the human and the organization at the heart of the complexity of organizations, cindynics explores, beyond technical characteristics, all categories of risk sources, as these also involve human, organizational, managerial, cultural and environmental aspects. These risk sources, which are generally more difficult to identify, receive much less attention.

It is from this scientific approach that the **MasKaDom project was developed to explore the gaps, deficiencies and dissonances existing in the different manufacturing designs of these types of masks.**

In this way, wearers will be able to gain greater confidence in their mutual protection.

E.4.3. *The cindynics approach*

Faced with the vagueness of contradictory recommendations, what can the general public do to build confidence in the preservation of their health by ensuring a reduction in the spread of the virus?

This is the essential reason that once again prompted the *Institut de la Maîtrise des Risques* to question the emergence of risk sources, which can be created by the lack of sharing of data, rules and models between all stakeholders in the same organization, and also of the purposes and values around the demand to wear masks and the processes of their manufacture and maintenance.

E.4.3.1. *The project's objectives*

The need for masks for non-sanitary use for the general public has become imperative and is rapidly growing in the context of collective protection: "I protect you, you protect me".

However, the effectiveness of homemade fabric masks depends on many factors, some of which have never been tested. As a result, these types of masks can pose a viral hazard to their users, especially after poorly controlled washing/drying cycles.

The aim of the project was therefore to provide the public with recommendations and advice for prototype masks and fabrics through performance indicators.

This project was created with great scientific rigor, which will also improve the confidence of the population in the quality and preservation of their health.

E.4.3.2. *The situation*

There are two ways of looking at the situation: the observed situation and the projected situation, in order to establish a comparison.

The first is the one practiced in France and is described above in the general context (see section E.4.1). The second is the one presented by the project's objective (see section E.4.3.1).

The actors selected for observation were mainly government authorities (both political and scientific) and the population.

E.4.3.3. *Systemic deficits*

This exploration at the global level sheds light on the **context** (cultural, managerial and organizational) in which the behavior of the various stakeholders has developed.

The main systemic deficits identified are discussed as follows:

CSD 1: *culture of infallibility*: no preventive measures were triggered in France before March 2020. However, on February 3, 2020, the WHO released the international Strategic Preparedness and Response Plan to help protect countries with fragile health systems[7]. But the scientific advice on Covid-19 was only implemented on March 11, 2020 by the Minister of Solidarity and Health, in order to inform the French public about the decisions in the management of the health situation related to the coronavirus.

At this stage, the example of Taiwan should be highlighted, because as early as January 5, this country set up a monitoring committee and identified Covid-19 as the origin of this pandemic.

We have already mentioned this general feeling of infallibility. French culture is not immune to it, which **translates into pathogenic aspects**. This particularly dangerous behavior completely removes the feeling of vulnerability at the risk of the safety of the entire population.

Moreover, it is amazing to recall that during the H5N1 flu epidemic, the preventive purchase of masks by the French Health Minister was seen as a waste of public money. This likely led to their subsequent unconscious disposal. And so France found itself without any masks when the pandemic started!

CSD 4: *navel-gazing culture:* in the face of the health crisis, the French executive chose to assert centrality, and even more so, that of verticality. "But while

7 See: https://www.who.int/fr/news-room/detail/27-04-2020-who-timeline---covid-19.

the health crisis, in many countries comparable to ours, provokes a reflex of sacred union around its leaders, in France, it has at times exacerbated and broadened political distrust: exacerbated, because the widespread feeling of having been the victim of a state lie on the issue of masks further damages the relationship between citizens and the government"[8].

Moreover, in this general context of mistrust of expertise (Expertise and Democracy, Dealing with Mistrust[9]), the failure of the state to provide barrier masks and the massive restriction of rights and freedoms by the declaration of a state of health emergency on March 23, 2020, have paradoxically aroused a desire for autonomy among citizens. They have taken hold of the resulting problem by starting to make masks themselves, since "it is better to have a self-made mask than nothing at all".

CSD 5: *priority given to production over safety aspects*: resignation when faced with organizational tasks is a fairly widespread evil that explains many malfunctions. Here, producing masks in an emergency, regardless of the method used, has been given priority over the safety of those who are going to wear them.

CSD 7: *lack of feedback*: the experiences gained during the numerous epidemics and pandemics have not been capitalized upon. France found itself in a state of unpreparedness, both for its leaders and for the healthcare personnel and the population with a lack of masks, gowns, protective gloves, respirators and so on.

CSD 10: *lack of preparation for crisis situations*: according to the elements reported in the context, the management of the crisis created by the Covid-19 was very chaotic in terms of the wearing of masks. And especially for this crisis, despite the existing regulations (white and blue plans), the health system was overwhelmed by the number of contamination and resuscitation cases. The public health service also made little use of private services.

The practice of just-in-time delivery in hospitals has become the norm, reducing their ability to deal with major health crises. Even the President of France expressed this unpreparedness in his speech on March 20: "What this pandemic reveals is that there are goods and services that must be placed outside the laws of the market"[10].

8 See: https://up-magazine.info/decryptages/analyses/59970-bilan-de-la-gestion-par-letat-du-covid-19-les-derives-de-la-verticalite/.

9 See: https://www.strategie.gouv.fr/sites/strategie.gouv.fr/files/atoms/files/fs-rapport-expertise-et-democratie-final-web-14-12-2018.pdf.

10 See: https://www.couches-infos.com/docs/ils_savaient.pdf.

All of these findings highlight numerous **gaps** in knowledge in the field of risk management and the sharing of this knowledge by the population.

E.4.3.4. *A look at the deficits*

This section highlights the inadequacies that are at the root of the impossibility of creating sharing and cohesion between stakeholders. To do this, the tool used was a questioning of the five cindynic dimensions that make it possible to sift through the inadequacies capable of "fabricating" risk sourcing that can lead to undesirable and feared events.

E.4.3.4.1. Data or facts

Some of them are mentioned in the paragraphs specifying the contexts (see sections E.4.1 and E.4.2.1). Other data can be added, such as those collected in the press: "The home-made cloth mask allows some of the large droplets to be filtered out, but their airtightness and ability to filter aerosols are not guaranteed". This mask can thus "bring" a certain protection, "limited", to a healthy person who would wear it in front of a contaminated person, notes Afnor, but **"this level of protection remains unknown. Beware, it is misleading to feel protected (or to think you are protected from others) from the virus with a simple fabric that does not meet standards"**[11].

The results of the tests performed by the DGA and concerning the filtration/permeability capacities of a list of fabrics and fabric complexes are publicly available, but in a form that is not easily understood by the general public.

All of the data available to the public is currently incomplete, fragmented and unreliable, and leads to risk prevention decisions that are at best arbitrary, and at worst ineffective or even dangerous.

Various publications have warned against the unreasonable use of cloth masks. Moreover, the profusion of data currently inhibits the use of truly significant data.

The study of Asian and European contaminated population curves pleads in favor of the adoption of masks by the population. But beware of selection bias, because these are the same countries that have implemented a policy of large-scale testing and population monitoring[12].

11 Le Monde – May 20, 2020.

12 See: https://phico.io/coronavirus.

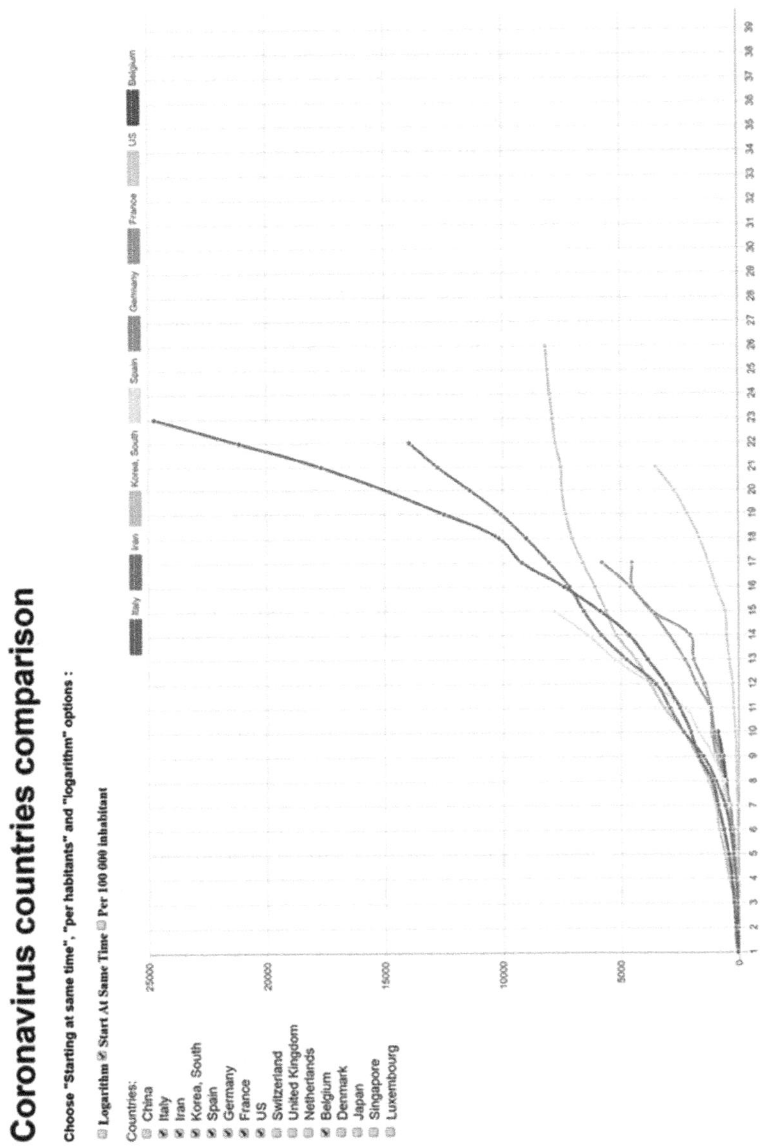

Figure E.3. *Comparison between countries generalizing or not the use of anti- projection masks. For a color version of this figure see, www.iste.co.uk/planchette/cindynics.zip*

According to the SFMC, the effectiveness of homemade cloth masks depends very much on the material used, the shape of the mask, its use and maintenance. Subject to continuing to strictly comply with government regulations on quarantine, social distancing and hygiene precautions, which must in no way be reduced, even partially, their use seems to complete the range of barrier measures recommended with the following reservations:

– have rapid efficiency tests to recommend their manufacture and use: five fabrics and five manufacturing methods could be rapidly tested;

– use these masks during authorized visits outside during lockdown or when lockdown is lifted when in the presence of a third party;

– knowing that the anti-projection masks protect the other side. **It is therefore necessary that both interlocutors wear a mask**: "I protect you, you protect me". Like in the game of ping-pong, both must participate. Wearing an anti-projection mask is a civic act;

– make sensible use of masks: in open spaces. In the absence of proximity it is not useful to wear the mask. **Its duration of effectiveness is four hours maximum**. It is thus necessary to use it economically;

– know its limits: it is not effective if it is wet and should not be worn if it hinders breathing;

– know how to remove it by taking it off by the elastics or straps;

– know how to maintain it: after use it is potentially contaminated; it must be placed in a dedicated container. It can be reused after **washing at 60° with detergent**. Additional cleanliness is ensured by ironing [JUL 20].

At this stage, only the regulatory aspects established by the DGA (filtration and breathability levels) and the AFNOR S76-001 specification are known for fabric masks. It would therefore be judicious and advisable to clearly define the content of the series of complementary tests to verify the best efficiency of the mask shapes and fabrics that will be recommended to the public.

In conclusion, the SFMC calls on the IMdR to quickly have efficacy tests available to develop recommendations for the manufacture of masks and their use, in order to restore the confidence of those who will wear these masks.

In addition to the breathability and filtration tests already defined by regulation, the questions to be asked to define the missing data that would allow this objective to be achieved concern the following areas:

a) The tests:

– for masks made at home, which types of masks and which types of fabric assembly are the most efficient?;

– characterize the different types of existing microorganisms in the ENT sphere, study their origin and their persistence on fabrics by using microbiological tests;

– check breathing capacity if the mask is blocked, as it may become difficult to breathe;

– evaluate the leak rate – which is a measure of the proportion of aerosol passing between the mask and the face – since the mask must fit snugly against the face to ensure a good seal;

– check the congruence according to the types of masks, as well as their adaptation to the morphology of different types of faces (child, adult, etc). This check is already carried out when wearing a mask for protection against asbestos or radioactive contamination.

b) Inter-comparability:

– between textiles, testing laboratories and **several types of coronaviruses.**

c) Washing:

– approximately 5% of the French population does not have a washing machine, and many people are now reluctant to the idea of increasing the temperature of their machine washes to 60 degrees[13]. How can we respond to these difficulties?;

– ANSM recommendations have been issued for a minimum 30-minute wash cycle at 60°. However, there is no public data concerning the wash time at 60° during a washing machine cycle for general public use;

– no test seems to have been carried out on the effectiveness of washing in the presence of Sars-CoV-2 according to the usual parameters of temperature, washing time, action of detergents and mechanical action;

– other checks do not seem to have been carried out on: the scenarios of use, transport and storage after use, the possible storage time between washing and reuse of the mask.

13 See: https://www.lci.fr/sante/coronavirus-covid-19-desinfectant-pour-le-linge-defroisseur-vapeur-seche-cheveux-que-valent-ces-techniques-contre-le-virus-2153690.html.

E.4.3.4.2. Models on which it is possible to rely on

The validation protocols for models used to test filtration and breathability are not disclosed. But are they based on the impact of sneezing and coughing episodes?

According to a document from the *Société Française de Médecine de Catastrophe*, it is mentioned that: "The emission of viruses is carried out in the form of droplets or aerosols (Pflüge droplets, Wells[14] *droplets nuclei*)".

They can remain airborne for up to six hours in a confined space[15].

The term droplet is used to refer to particles with a diameter greater than 5 μm, that fall rapidly to the ground by gravity, and are therefore only transmitted over a limited distance (about 1 m). Droplets projected by sneezing or coughing efforts are projected up to 8.2 meters [BOU 20]. In contrast, "droplet cores" of particles with a diameter of less than 5 μm can remain suspended in the air for long periods of time, allowing them to be transmitted over distances of more than 1 meter [STE 04].

Other models will therefore be developed to evaluate the additional tests cited in the "data" paragraph.

E.4.3.4.3. Laws and regulations (see Chapter 1, section 1.3)

The evolution of the doctrine of mask wearing (see section E.4.1.2) has led to a loss of interest and trust in the information available on masks.

E.4.3.4.4. The values that guide us (see Chapter 1, section 1.3)

The term **values** refers to attributes and perceptions that a person shares with members of their social or cultural group. Types of sociological values include moral and ethical, ideological (political), spiritual (religion), beliefs, for example, ecological, or aesthetic, open, individual and collective.

In public perception, the values displayed regarding the obligation to wear a mask have been interpreted as a **swing between the priority of health and the priority of the economy**. It would be useful to **clarify the choice in order to ensure support.**

Some other values such as autonomy, security and compliance were hardly shared. Is this related to the message that is being sent to the public and which is part of what cindynics sees as *dissonance,* which is also a risk source?

---

14 World Health Organization, 2009: https://www.ncbi.nlm.nih.gov/books/NBK143281/.

15 See: http://campus.cerimes.fr/microbiologie/enseignement/microbiologie_3/site/html/3.html.

E.4.3.4.5. Purposes or goals (see Chapter 1, section 1.3)

Here, the term purpose translates the *raison d'être* of the barrier masks, the goals sought.

In the situation of a shortage of FFP2 and surgical masks, it is therefore a question of providing the public with alternative masks to protect themselves, in addition to barrier gestures and rules of distancing, but also to guide their manufacture as well as their maintenance, with all of the scientific rigors necessary for this population to find or regain confidence in the recommended masks.

It seems that very different purposes, for example of an economic nature, can be observed according to certain stakeholders, without this being clarified, due to the lack of an open discussion.

E.4.3.5. *A look at dissonance*

Here are some examples:

– "Scientific experts have expressed divergent opinions and have frequently changed their views. The terrifying predictions of mathematical models have not come true. Major journals have published results of contradictory experiments, and have sometimes even retracted them. Science may well be discredited as a result".

However, the lack of consensus among experts is rather a symptom of a healthy scientific debate. Knowledge is the fruit of a continuous debate on the interpretation of facts, themselves established with more or less precision. Different researchers advocate different ideas [LAN 20].

This type of dissonance is above all a producer of progress and should be considered in a positive way (see Chapter 3, section 3.2.2), even though it disturbs those who think that the scientific word should be considered a certainty:

– other polemics between scientists and laboratories have unfortunately produced misunderstandings among the population, such as the one concerning the hydroxychloroquine molecule;

– some messages from the public authorities have been contradictory. For, how can we understand that schools and restaurants must remain closed, while at the same time, people are asked to go and vote in polling stations where, by definition, other people are present?;

– the reactions of some citizens who, although informed of the seriousness of the situation and of the preventive measures to be adopted, will continue to hold each other close, shake hands, kiss each other, … for what reasons? Not to feel constrained? Bravado? Ignorance? Or for other reasons that have not yet been identified?;

– the contradictory recommendations established for the making of masks at home.

These other examples of dissonance have only reinforced the mistrust between experts, governments and the population. This sense of mistrust represents a major risk source.

In conclusion, this study, based on the state-of-the-art, shows that several pathogenic elements have emerged in current approaches, and that these vulnerabilities favor the creation of dangerous situations conducive to the emergence of a resumption of the pandemic. **According to the European Centre for Disease Prevention and Control, the question is not whether there will be a new wave, but "when and to what extent"[16].**

It therefore seems essential to question the validity of the conceptual model of SARS-CoV-2 transmission, to broaden the typology of tests, taking into account the shortcomings already mentioned, to disseminate effective tutorials for the manufacture of masks for general public use and to strive for appropriate communication to (re)gain the trust of the population.

The amplitude of the verifications proposed by the MasKaDom project meets the requirements of the cindynics approach, requirements which will make it possible to reduce many uncertainties, as well as the strong emotional charge currently conveyed.

This project should also be supported by communication adapted to the desired gain in trust.

E.5. An interdisciplinary look at all the examples

The four examples studied clearly illustrate the interest, in the presence of complex situations, of choosing a cindynics approach adapted to their context and rich in lessons.

Each of the examples has been chosen for its own specificity:

– The case of Bhopal encompasses many factors – socio-demographic, regulatory, organizational, economic, physical, institutional and even political-administrative – at the level of a country, India. In addition, the decisions on

16 L'Express from May 21, 2020 – 6:23 PM, updated at 8:36 PM.

technical choices and operating methods of the pesticide manufacturing plant were made by another country (the USA). This resulted in risk sources specific to these choices (political, economic, cultural and linguistic) that were amplified by a cascade of slow transformations between 1957 and 1984. The elements of the drama then gradually came together in such a way that all it took was a match (high-pressure cleaning by insufficiently trained personnel) for the disaster to occur.

– As far as the Queen Mary is concerned, the construction of a gangway was "insignificant" compared to the construction of this huge liner. And yet, it was this "tiny" piece of equipment that caused a catastrophe resulting in 16 deaths and 30 being injured. In this case, the activities took place in the same country, in the same place, and a "small phenomenon" disrupted the organization of a huge shipyard: the principle of integrating the public from outside the company to visit the ship before it was delivered. On the contrary, the urgency made the company forget the essentials (the physical dimensioning, as well as compliance with procedures when placing orders for gangways) and the existence of subcontracting taking place on the site of the shipyard itself, which did not benefit from a proper formalization of the sharing of responsibilities between project owner and project manager.

– Deepwater was an offshore oil extraction base in the deep waters of the Gulf of Mexico, off the coast of Louisiana. Its location was close to very valuable ecosystems and posed ecological, economic and political problems.

The BP disaster shows the imprudence of considering that a small risk is equivalent to no risk. Indeed, probability calculations were based on previous statistics, whereas this in-depth operation was unprecedented. This overconfidence also explains the lack of preparation for the crisis.

Another peculiarity of this disaster lies in the nature of its disastrous consequences: in the face of BP's involvement, which appeared to be one of the most advanced in terms of societal responsibility, the ecological damage has reached such a severity that today no one is able to assess the environmental after-effects on the deep ecosystems. In this case, there is also no physical understanding of the phenomena at the time of the accident.

This event could change the rules of the game of oil environmental regulation in sensitive areas, and become one of the reasons for accelerating the transition to new energies:

– With regard to the Covid-19 approach and the issue of masks, this is a study conducted *a priori*, with the aim of providing recommendations to improve the quality of homemade cloth masks.

This study, carried out after the work of the GTR, aims to show that cindynic approaches can be perfectly used before the occurrence of an accident. The power of cindynics applied to actors' behaviors provides a specific view of the complexity of situations, which no classical risk study can achieve. Thus, it is possible to evaluate the downward spirals that inevitably occur through slow transformations.

It is also interesting to note that the robustness of cindynics allows a variety of representations of hyperspaces and matrices, without affecting the results.

In addition, these examples highlight common points in the study of activity situations:

– the importance of **understanding their contexts** and enriching them as much as possible with new desk research;

– the interest of **deepening or readjusting the dimensions of the horizons of the situations** (appearance of new actors whose involvement proves to be important afterwards, etc.);

– the attention to be given to **slow transformations, changes in pace or nature,** moments considered critical in the activities (example of process re-commissioning, ageing of installations);

– the management of interdisciplinary as well as vertical, intra and inter-organizational interactions, because **interface problems cause** organizations to lose resources and can be the source of vagueness, ambiguity and so on;

– the need to not only focus on the occurrence of a dreaded event. For **as long as the study of accidents and disasters focuses on the hazard, "human error" will be favored and will erase the consideration of pathogenic effects** that are insidiously generated during organizational changes;

– the originality of a cindynics study that allows for searching for the root causes that have led to major dysfunctions or accidents, because if "latent errors" [REA 13] are not detected and treated, the "human body" that the organization represents will not be completely cured and relapse will always be possible.

Appendix 1

Current Risk Management and its Shortcomings

A1.1. The main principles of the current process

The principles and guidelines of risk management are established by the NF ISO 31000:2018 standard, and the practical aspects of risk assessment techniques are presented by the IEC/ISO 31010:2009 standard. It states that:

> Risk analysis consists of determining the consequences and probabilities for identified risk events, taking into account the presence (or not) and the effectiveness of any existing controls. The consequences and probabilities are then combined to determine a level of risk.

This operating process, while it has brought real progress in the reduction of adverse events, unfortunately has some shortcomings in relation to current scientific advances. In the following, we will highlight five shortcomings in risk analysis and risk management which are still frequently encountered. These shortcomings allow us to provide explanatory elements to justify the limits, biases and solutions that can be used to compensate for them, thus showing the value of integrating the scientific advances of the second half of the 20th century and the beginning of the 21st century into traditional risk management approaches.

Indeed, since the beginning of the 20th century, the reality of the world has changed considerably thanks to two fundamental discoveries: **quantum theory**, which has brought to light the existence of invisible aspects of matter, and **complexity**. It should be added that during the same period, numerous advances in the social sciences have profoundly changed the way we look at the behaviors of individuals evolving within any social process.

From these three advances (complexity, quantum theory and social psychology), three essential ideas for the management of major risks have emerged:

The Cartesian approach, perfectly adapted to the study of stable systems constituted by a limited number of elements with linear interactions (i.e. that can be described by continuous and additive mathematical laws), is **no longer appropriate as soon as we are faced with organized complexity** such as that encountered in large biological, economic and social systems.

All **perceived** reality is our own and can be described, but it does not represent reality in itself, because **there are other inaccessible elements (ultrasound, infrared) that can be described by applying, for example, the method of relativized conceptualization (MCR).**

If we move beyond the product stage and into the life of an organization that produces it, in this case, a collective process leading to a common goal, the **behaviors and attitudes of all community members** are also important to study.

A1.2. Limitations of the current process and solutions

– **The first shortcoming** lies in the unique use of analytical methods that rely on Cartesian reasoning to assess possible system failures and their consequences. Each piece of equipment is subdivided into as many parts as necessary in order to analyze its operation with tools such as FMEA (Failure Mode and Effects Analysis), HAZOP (Hazard and Operability Study) or HACCP (Hazard Analysis and Critical Control Points), etc. This method can be viable if it is limited to one piece of equipment.

However, when it is transposed to the level of an organization, it is not suitable, because an organization is like a complex system with internal and external actors interacting with each other. Therefore, all the characteristics of complexity apply (interactions, emergence, loops, levels of organization, types of information). Thus, the notion of interaction between the parties goes beyond any simple cause and effect relationship and the behavior of each party cannot explain the behavior of the organization when considered as a system, taken as a whole.

Consequently, any organization, being endowed with a purpose and composed of parts, must also be considered as a whole, and a new form of danger would be to want to "dissect" the system in a purely analytical way.

Therefore, it is essential to integrate a more systemic approach to dealing with hazard and risk management in this case.

– **The second shortcoming** is based on the notion of risk by excluding the hazard at its source. Indeed, unconsciously relying on the presumption that all existing hazards are now known and cataloged by numerous experiments, hazard studies are replaced by risk studies. However, hazards and risks are not the same. A hazard is the capacity to cause damage, as in the case of a toxic product, whereas risk only exists if an issue (people, equipment, etc.) is confronted with one or several hazard(s), either by undergoing it (risk suffered), or by deciding on it by a voluntary act (risk taken) and that a triggering event is capable of materializing the hazard(s).

It is therefore necessary to restructure the notion of danger, because we cannot be certain about the completeness of our knowledge of existing hazards. The shortcomings developed later will make this clear.

– **The third shortcoming** relates to the "event-driven" approach to risk, from the perspective of scenario combinations. In addition to the fact that it reinforces the evaluation of the risk by quantitative parameters (likelihood and severity), it directs suspicion towards the human operator at the origin of the event and ignores the inevitable internal evolutions that may be at the origin of new dangers. This approach, called the "human factor", is very reductive because it only seeks to illustrate human reliability and human–equipment relations. After reading a certain number of bibliographical references on the subject and study reports, we have noticed that these approaches were mainly focused either on the appropriate way to assimilate humans with the machine, whose reliability can be calculated, or on the active errors associated with the performance of front-line operators. This shortcoming is illustrated by James Reason's conclusion in his book *Human Error* (2013); see Chapter 7 for further discussion.

The studies on the genesis dynamics of disasters and diffuse accidents presented in the book *L'archipel du danger* [KER 91] have clearly shown the impact of organizational and societal constraints on the actors of the system, generating behaviors in each of the stakeholders (actors or groups of actors) composing an organization. Each stakeholder is caught up in a set of constraints and plays an important role, regardless of their level of responsibility.

This psycho-sociological approach goes far beyond the framework of "human factor" studies, focused on the last link in the operational chain where the human factor is supposed to make active errors. Indeed, the foreword to the latest edition of Reason's book mentions this problem:

> Although the notion of human error is popular, it nevertheless contains some limitations that James Reason highlights in the analysis of accidents at the end of the book. One of these limitations is that the cause of accidents is, at least in part, human. This is a limitation because the crude term "human error" says nothing about the managerial decisions that created the conditions for its occurrence. Nor does this crude term say anything about the pathogenic contextual factors that acted on the operator at the origin of the human error [REA 13].

Let us therefore learn to make the concept of "human factors" evolve towards that of "behavioral factors" of all the actors.

– **The fourth shortcoming is that the risk is evaluated only through a mathematical expectation**, in order to qualify the risks according to the probabilities of the triggering events and the severity of the consequential damages. In general, this gives indications of the internal workings of the organization and what it is prepared to accept. Because of the complexity of the problem, this type of assessment only very rarely integrates the effects of internal and external interactions. Moreover, this process excludes any event whose description is not quantifiable, whereas the existence of a human collective within an organization shapes a set of more or less shared values, beliefs and representations. Qualitative aspects such as values and beliefs strongly influence the actions of individuals and can create differences in beliefs, perceptions, purposes and values. These conflicts can lead to forms of paralysis or blockage, which are risk sources that have not been taken into account before now.

What probabilities could then be associated with these qualitative factors?

Combining these qualitative aspects with the quantitative assessments represents a significant qualitative leap forward in addressing this fourth limitation.

– **The fifth shortcoming** is due to a sensory perception of danger. Our brains have mental tools that allow us to evaluate, in a way that is considered to be scientific, a large number of the dangers that our senses perceive and therefore the associated risks. Indeed, these phenomena are observable in their context and their harmful power.

Quantum theory has taught us that knowledge of reality only corresponds to what seems observable to us through our senses and our measuring devices. The contribution of the discoveries of quantum physics has made it possible to show the existence of invisible aspects of matter, thus dispelling the idea that the only sources of danger are those accessible to our senses. Indeed, we know that humans are physiologically incapable of perceiving all the warning signals. Let us take the example of earthquakes. On several occasions and well before the appearance of the consequences linked to natural disaster phenomena, it has been observed that animals behave in a way that seems strange. However, we already know that we are insensitive to ultra-sound, ultra-violet, infrared and magnetic fields. Animals are likely to sense warning signals that make them flee risk zones (such as in the tsunami of 2014). Another example of non-sensory dangers lies in the interplay of the actors involved in a company where everyone has their own values and interests at heart. These discrepancies can generate a major source of incoherence and thus create dangerous situations.

However, if there are sources of danger that are not perceptible to our senses, what forms can they take and how can we deal with them?

These questions are central (see section 2.3.1 for further discussion).

The fifth limitation encourages us to go beyond our sensory approach to identify unidentifiable or latent dangers.

Appendix 2

Notions of Interaction and Complexity

A2.1. Preamble

"Life is a relationship among molecules and not a property of any molecules" noted the eminent American chemist and physicist Linus Pauling.

Morin, in *La méthode, Volume 1: La Nature de la nature*, adds, "**the organization in systems produces unknown qualities or properties of parts conceived in isolation: emergences**".

These types of relationships are called **interactions**, and **emergences** are properties of complexity.

A2.2. Notions of interactions

An interaction is a **reciprocal action or influence between two or more elements** of a system or different systems. It has a change effect that participates in the evolution of the system:

– either on the scale of the integration of elements (for example, pollinating insects that co-evolve with flowering plants);

– or by creating new properties through the emergence effect (for example, mayonnaise).

This notion of interaction can be used in various disciplines (biology, chemistry, mathematics, physics, medicine and social sciences).

It is therefore inevitably used in the context of risk control, where dreaded events "emerge" from the interactions between the various elements that constitute our socio-technical systems.

However, in traditional risk management, instead, we have tended to favor the study of causes linked to technical interactions whose actions retain a proportionality between them (**linear dynamics**).

However, as our socio-technical systems are an assembly of technical, human and social elements, it seems essential to integrate all aspects relating to the field of **social interactions. These refer to all reciprocal actions between two or more individuals during which information is shared**; for example, an operator and a maintenance man will interact by discussing in a prior context known to them: the exchange of information.

Interaction is said to be social because not only does it **produce meaning**, but also because it takes place in a **context** that **influences** the actions of each person.

Interactions are *verbal* or *non-verbal* (gestures, looks, attitudes, etc.) and can induce:

– *cooperation*: participation, adaptation, integration, emulation, cooperation, etc.;

– *conflict*: struggle, rivalry, segregation, discrimination, insults, etc.;

– *ambivalence*: competition, rivalry.

As a result, the integration of human and social aspects leads us, through the interplay of social interactions, into the universe of **complexity** where the nature of the interactions made up of positive or negative feedback loops creates a **nonlinear dynamic**, that is, there is no proportionality between effects and the causes that underlie them, nor is there any additivity between the causes and the effects.

A2.3. Notions of complexity

Starting from the etymological sense, complexity is presented in two dimensions, that of "**entirety**" and that of "**interaction**".

"**Entirety**" refers to the first etymology of the word complex, which is "*complexus*", a verb meaning "to embrace, to encompass". Seen from this angle, **complexity is then the fact of a reality that brings together, in solidarity, several different elements.**

Together with the second etymology "**interaction**", already mentioned in the previous paragraph, this concept refers here to the idea of things "woven together" which **emphasizes more strongly the link, the relationship between the elements**. Jean-Claude Ligeron in "*Le Cercle des fiabilistes disparus*" cites, to explain what complexity is, the "presence of a large number of independent interacting elements" [LIG 06].

This concept of interaction, one of the richest in the system, complements that of entirety **because it deals with the complexity at the elementary level of each relationship between the constituents of the system taken in pairs**. Initially borrowed from mechanics, where interaction is then reduced to **a play of forces**, the relationship between constituents is most often translated in complex systems by a **relationship of influence or exchange** relating to flows of matter and energy as well as information.

As already indicated in the concept of feedback, the notion of interaction goes far beyond the simple *cause and effect relationship* that dominates classical science. **Knowing the nature and form of the interaction is more important than knowing the nature of each component of the system.**

However, these two elements of meaning alone are not sufficient to capture what is currently referred to as **complexity**. Indeed, several authors have already stressed that, in terms of complexity, the characteristic is not the quantity of elements taken into consideration, nor the multiplicity of relations they have together, but rather **unpredictability**!

The complexity of a system is due to at least three factors:

– **the high degree of organization**: to illustrate complexity, Laborit evoked the number of levels of organization by specifying: "that atoms will come together to form a molecular assembly, that these molecules will constitute the elements of a new, enzymatic assembly, that these molecular groups will allow an enzymatic reaction to take place by coming together [...] until they form a cell assembly, a cell, [...] until they structure a human organism";

– **uncertainty**: as an image of unpredictability characterized by the "butterfly effect";

– the difficulty, if not **impossibility, of identifying** all the elements and understanding all the relationships involved.

Consequently, the laws used to describe a system cannot be purely deterministic.

Other characteristics underline the peculiarities of complex systems:

– **information**: it is permanently involved in exchanges between and within systems, alongside the other two fundamental flows of matter and energy. Laborit identifies:

- *structuring information*, constituting a memory whose form persists over time (included in the system's memories, for example, the DNA strands of the chromosome for a living cell or, more simply, language, books, monuments, etc.);

- *circulating information* (to be treated as a simple perishable flow). "Between each organization level, there is circulating information so that, from one level of organization to another, the whole can function harmoniously and, above all, maintain its own structure".

The circulating information is of significant interest since each level of the organization receives this type of information from the lower levels, allowing it to maintain the overall structure of the system as well as the integrity of each organization level.

– **the loops**.

Exchanges can take the form of all types of energy, information, etc., even constraint. In addition, interactions can simultaneously take place with several subsystems and loops, such as the thermoregulation mechanism of the human body. The study of loops also illustrates uncertainty and unpredictability, which breaks the principle of linear causality.

Appendix 3

The Grounded Theorization Method

This approach, which is an adaptation of grounded theory developed by Glaser and Strauss in 1967, is dedicated to "qualitative analyses of raw data collected from field notes, interview transcripts, various documents, etc." [PAI 94].

Based on these data, their analysis consists of a "long and meticulous interpretation activity, passing through a series of precise procedures, a series of reflections, questioning, discoveries and constructions that are ever more enlightening and better integrated".

"Grounded theorizing analysis [...] is therefore an iterative process of progressive theorization of a phenomenon, [...] i.e. its evolution is neither expected nor linked to the number of times a word or a proposition appears in the data. Thus, it corresponds neither to the logic of applying a pre-constructed thematic grid nor to that of counting and correlating mutually exclusive categories. In fact, grounded theorizing analysis *is not just content analysis*; it is much more accurately equivalent to an act of conceptualization.

It is then a question of "identifying the meaning of an event, [of] linking various elements of a situation in an explanatory diagram, [of] renewing the understanding of a phenomenon by putting it in a different light".

"Grounded theorization analysis is based on data collected through qualitative research, a major feature of which is the *simultaneous collection and analysis,* at least in the early stages [...]. The analysis therefore begins at the same time as data collection. Each series of two or three interviews or observation periods must be followed by the transcription of the data, followed by the analysis (which will be rather basic at first, but will become more complex as time goes on [...]). This analysis will, in turn, help guide the interviews or observations, etc. This means that these can take place virtually up to the stage of preparing the final report".

Regardless of the level of theorization achieved, "the result will have to be *grounded* firmly in the empirical data collected, [constantly ensuring] that it corresponds closely to the data".

A3.1. The five steps of data analysis by grounded theorization (source: Wikipedia)

– **Codification**

The aim is to summarize, in the margins, the verbatim statements made by the respondents, keeping as close as possible to what they said.

– **Categorization**

This step consists of re-coding the corpus with a richer tool: the category. The codes or labels placed on the verbatim summaries are grouped together, merged by "naming the phenomena and events that emerge from the data in a richer and more encompassing manner" [PAI 94]. The question to be asked is: "What phenomenon am I facing?".

The desired outcome is "a word or phrase designating, at a relatively high level of abstraction, a cultural, social or psychological phenomenon as perceived in a body of data" [PAI 96, p. 186].

– **Putting people in touch**

The researcher will establish links between the different categories defined in the previous step. "With this linkage, [one] begins to access the event, the explanation, the story" [PAI 94]. This stage can also lead the researcher to rework, rethink and modify the categories decided upon during categorization.

– **Integration**

Grounded theorization brings about the problem that is never posed *a priori*. At this point in the analysis of the field data, the researcher asks themselves what the problem of their field survey is "in the end" [PAI 94].

– **Modeling**

During this stage, the researcher creates a diagram that reveals the dynamics of the phenomenon they are studying.

Appendix 4

Notions of Quantum Theory

The break with classical theories

The quantum theories born at the beginning of the 20th century marked a complete break with the physical principles accepted in the 19th century (Newtonian and electromagnetic theories). Through the study of "microsystem states", these new theories came to describe the structure of atoms and molecules and explain that matter is not only composed of visible elements but also has many *invisible aspects*. They have thus constituted a "conceptual change. For the first time, both pure reason and what is believed to be 'common sense' are being challenged by experimental facts. We must build a new way of thinking about reality, a new logic" [BAS 02].

Do our senses perceive reality?

Previously, knowing an object was being able to say what it is. To recognize a certain table, for example, was to be able to describe it in shape and appearance (length, width, height, constituent materials, color). The knowledge consisted of a **description** of an object, that is, a translation transmissible by a language of reality as it was approached. "All our language, all logic and classical probabilistic thinking are based on this postulate, more or less implicitly" [MUG 77].

Now, quantum theory shatters this conception of knowledge acquisition, for it shows that the world of the infinitely small behaves very differently from our macroscopic universe. Every observer influences the state of the microscopic system observed with their instruments and according to the circumstances of the observation. **The principle of the observer's separation from the observed also "renders inoperative the concept of reality in itself"** [BAQ 07, p. 59].

Consequently – and this notion is now corroborated by the neurosciences – our experience of reality does not show us its completeness. We grasp it only through the modification of our senses and the prism of our perception. **The reality we perceive is our reality, not reality as it is in itself.**

How can we conceptualize, know, study and describe (i.e. qualify) unknown objects or states?

A new way of thinking about reality

At the beginning of any conceptualization chain, there are physical operational acts of delimitating physical entities, such as taking a sample for analysis. To be able to qualify an object, it is therefore first necessary to delimit the subject of study. However, access to knowledge is not guaranteed by this act of delimitation, because it is necessary to associate a certain number of examinations that generate qualifications or attributes (shape, color, etc.) for this access to be possible. Applying this to the macroscopic world, our usual logic postulates the existence of pre-existing attributes. However, in the microscopic world, the qualification of objects is unknown *a priori* and can only be obtained by acts of qualification, which are moreover only potentialities (or probabilities) of characterization. These potentialities highlight the character of relativity that these acts of qualification impart to any description. Consequently, **any method of conceptualization becomes relativized.**

In 1982, the physicist and philosopher of sciences Mioara Mugur-Schächter laid the foundations of a method of relativized conceptualization (MRC) that she gradually expanded on in her book *Sur le tissage des connaissances*, published in 2006. "In our opinion, this work constitutes a revolution in the way we represent the processes of knowledge acquisition and, consequently, the 'reality' or 'world' that is the object of this knowledge"[1].

In this work, Mugur-Schätchter applies a rigorous method which consists of an operation of generating an object of study, then forming a view composed of a set of views corresponding to certain qualifying properties (which she calls aspects) of the generated object.

With this method, it is possible to invent projects of objects "which are not directly perceptible to the human senses: the invisible, the inaudible, the impalpable, the infinitely large or the infinitely small, all of these many worlds requiring the use

1 Additions to the book *Pour un principe matérialiste fort* by J.P. Baquiast [BAQ 07].

of traces of phenomena that is absolutely impossible for us to grasp directly" [MUG 06].

The main construction stages of the MRC method

Let us briefly mention the main steps that are essential to the construction of knowledge according to the MRC [MUG 02]:

Consciousness functioning: the existence of a human observer, endowed with a conscience and a brain capable of elaborating goals to which a strategy can be put at the service of. This observer has the ability to construct representations of the world around them from the sensory data they collect.

Reality: another postulate, beyond the constructions by which this observer represents the world, there is something else that it is impossible to describe objectively.

Generator of Object–Entity and the Object–Entity [MUG 02]: thus generated, this is a mechanism allowing the Consciousness–Functioning, within the framework of the defined goals and strategies, to create an observable from which measurements can be made.

Qualificators: these are different points of view (length, width, color, speed) by which the object–entity created can be described. There is only one qualification per measure and this is not repeatable since generally the object–entity changes because of Heisenberg's principle of indeterminacy. The multiplication of the qualifications (or aspects) makes it possible to form a *view* composed of a set of *views–aspects* of the generated object–entity.

The *Frame–Principle*: corresponds to the space–time framework in which we decide to observe the object–entity in order to situate it.

By the multiplication of qualifications, this method results in a "relativized description" of the object–entity. It is thus no longer a question of simply detecting pre-existing properties of an object. Since there is no scientific knowledge without descriptions and since the "reality in itself" is deemed to be non-absolute, it is rather a question of applying a methodology to produce descriptions. In this approach, **knowledge becomes a construct that is enslaved to a given goal**.

Thus, thanks to this MRC method, it has become possible to conceptualize and describe unknown objects or states. As a result, risk sources not perceived by our senses can become accessible for description.

Appendix 5

Summary of CSDs

Deficit Area	CSD Number	Cindynics Systemic Deficit Designation
Cultural (4)	CSD 1	Culture of infallibility
	CSD 2	Culture of oversimplification
	CSD 3	Non-communication culture
	CSD 4	Navel-gazing culture (lack of communication)
Organizational (2)	CSD 5	Priority given to production over safety aspects
	CSD 6	Dilution of responsibilities
Managerial (4)	CSD 7	Lack of feedback
	CSD 8	Lack of a procedure with a cindynic dimension
	CSD 9	Lack or inadequacy of risk management knowledge
	CSD 10	Lack of preparedness for crisis situations

Table A5.1. *Summary of CSDs*

Appendix 6

Archeocindynic Study

René Baeli was the last director of the École d'ingénieurs de Marseille before its merger with Centrale Marseille. Knight of the French Ordre des Palmes Académiques[1], he was president of the scientific council of the Institut Méditerranéen des Cindyniques and vice president of the Association méditerranéenne pour l'étude des missions publiques de protection civile. For many years, he also provided scientific assistance to the Bouches du Rhône Departmental Fire and Rescue Service (SDIS13).

He was introduced to cindynics by Georges Yves Kervern and introduced the teaching of cindynics into the School's curriculum.

At the time, he had asked the engineering students to make a cindynic study of a disaster or accident of their choice. Some students had chosen to study historical or even legendary facts…

He was thus able to build up a database over several years and to gather these historical analyses in *Les Archéocindyniques* which applies the principles of cindynics to 16 events that have marked French history.

With the author's permission, we have extracted five events which we reproduce here, in addition to a brief reminder of the facts and the analysis of "systemic cindynogenic deficits".

The book has become a pedagogical objective showing that the methods associated with the science of danger can very well be applied to feedback of distant events, the analysis of which is always instructive.

---

1 A national order bestowed by France on distinguished academics and teachers for their services to education.

Analysis of historical facts

A6.1. Alesia: 53 BCE

Background

In 53 BCE, Caesar thought he had definitively subdued Gaul. But a young Arverne chieftain, Vercingetorix, succeeded in convincing the other Gallic tribes to join him to form a great army and free Gaul from the Roman yoke. His strategy was never to fight on open ground, but rather to harass the Roman army and abandon the cities when Caesar approached them, ransacking everything so that the Roman army could not get supplies.

In spite of the advice of Vercingetorix, the inhabitants of Avaricum refused to abandon their city which was taken and destroyed. But Caesar was forced to divide his legions into two groups to extinguish the hotbeds of rebellion. He tried to take Gergovia, one of the largest Gallic cities. Vercingetorix observed the military organization of the Romans well. Over several nights, the Gauls harassed the enemy guards. Litaviccos joined Gergovia with a troop of a thousand men. The cavalry, a strong point of the Gauls, drove the legion in and caused 700 deaths. Caesar was obliged to decamp. But his legions took Lutetia and Bibracte. Vercingetorix wanted to annihilate the Roman armies at all costs before they left, for he knew that they would return in even greater numbers. The Gallic cavalrymen constantly attacked the advancing legion. But the German horsemen recruited by the Romans counterattacked and made the Gallic cavalry retreat.

Vercingetorix chose to take refuge in Alesia, a stronghold reputed to be impregnable, and to wait for Caesar to fight him in a place he knew and had chosen. He sent part of his cavalry to seek reinforcement with the idea of taking the Romans by surprise by surrounding them, while they themselves lay siege to Alesia. After 30 days of siege, the battle of Alesia began: 70,000 Romans against 40,000 Gauls. (These figures are very controversial, Caesar in the Gallic War quoted 100,000 Gauls.) Inside Alesia, Vercingetorix kept only the elite of his army to save food. The Romans built immense fortifications. Gallic reinforcements were slow to arrive.

The battle, which lasted several days, began with a cavalry battle. It ended in failure. The relief army arrived and attacked at night from the outside at the same time as the besieged from the inside, but they failed to join up and break through the Roman fortifications. The next day, at noon, a surprise attack on a Roman weak point on the North Mountain nearly succeeded. Victory was within reach. It was then that Caesar, conscious of what was at stake, came himself, risking his life to galvanize his troops. The Gauls were pushed back and during their retreat, the German horsemen massacred them. Frightened, the survivors of the relief army fled and returned home, leaving Alesia under siege. Vercingetorix knew that, alone in Alesia, he would not be able to win, so he surrendered to Caesar. The survivors were given as slaves to the Roman soldiers. Intelligently, to divide the Gallic people, Caesar freed the Aedui and the Arvernians.

Deficits

Cultural

Infallibility

The Gauls believed themselves to be very superior in terms of cavalry. They were convinced that they could defeat Caesar weakened by the Aedui and the Narbonnais. After the victory of Gergovia, Vercingetorix thought he could do the same battle again. He chose Alesia, a hill reputed to be impregnable. The Gallic archers were reputed to be more efficient than those of the Romans.

Oversimplification

If Caesar abandoned Gaul, he would return with more troops, so he needed to defeat him at that moment in time. The Gauls, because they had chosen the place of confrontation, thought that they would be in a dominant position.

They did not analyze that Gergovia's victory was due more to the disobedience of the centurions than to their own superiority. If the Gauls had defeated Caesar, it was because the Gauls were the strongest and therefore could beat Caesar again.

Navel-gazing

Vercingetorix succeeded in uniting the Gallic forces but this was only superficial; each tribe retained its own interests, its own feelings towards the young chieftain and their own relations towards the Romans.

Lack of communication

Alesia's troops had no news of the reinforcements that were due to arrive. They were unable to get any information about the fortifications: investment, traps, sharp peaks, etc.

The troops from the outside were struggling, despite the trumpets and shouts, to know where to attack in order to torment the soldiers from the inside.

Organizational

Productivism

Food was beginning to run out in besieged Alesia. The Gauls were impatient. Fighting began before the arrival of reinforcements. The Gauls were valiant but were ungovernable when they felt they could win or when they fled in terror.

Dilution of responsibilities

The Gallic army, a patchwork of tribes, was very undisciplined. Vercingetorix had to convince each small chief to have a command unit. Many small acts of treason informed Caesar and weakened the Gauls.

Managerial

Lack of feedback

Vercingetorix did not take into account the superiority of the cavalry of the Germans, which was already tragically experienced. He did not understand that the victory of Gergovia was due to a scramble by the Roman soldiers who did not follow Caesar's instructions.

No cindynic procedures

The search for reinforcements was improvised. The withdrawal of the cavalry was done in the greatest disorder. The management of rationing made many people dissatisfied, especially the inhabitants of Alesia.

No cindynic training

N/A.

No preparation for crisis situations

Vercingetorix very poorly managed the call for reinforcements and the distribution of food during the siege.

A6.2. The 1720 plague of Marseille

On May 25, 1720, a ship, the *Grand-Saint-Antoine*, entered the port of Marseille, France. It brought back from Syria a stowaway, the plague.

In two months, the city of Marseille lost half of its 100,000 inhabitants and across the whole region, the plague killed no less than 220,000 people!

The French of the "*Age of Enlightenment*", who lived in the recklessness of the Regency (King Louis XV was 10 years old at the time) and believed they were safe from the great epidemics, had to achieve severe prevention in a catastrophe.

The plague, whose name comes from the Latin *pestis*, was only identified in 1894 by the physician Alexandre Yersin. It comes from a very resistant microbe which bears the name of its discoverer: the *bacillus of Yersin*. It exists in its natural state in certain Asian rodents and can be transmitted via fleas to rats and, from there, to humans. The flea in question is repelled by the smell of sheep and horses, hence why shepherds and grooms were safe from the disease.

Europeans initially believed that the miasmas of the plague were spread by air. Therefore, when the epidemic reached a city, they were in no hurry to flee. The poet Boccaccio recounts this in *The Decameron,* his collection of tales written after Florence was hit by the Great Plague of 1347. Flight is the worst attitude there is, because it has the effect of accelerating the spread of the epidemic.

The *Grand-Saint-Antoine* left Marseille on July 22, 1719 and reached the ports of call or ports of the Levant. At that time, the plague was raging in Syria.

A Turkish passenger embarked in Tripoli on April 3, 1720 and died two days later. Then, on the way back, the sailing ship lost seven sailors and the ship's surgeon in succession. An eighth sailor fell ill shortly before arriving in Livorno, Italy. Each time, good reasons were found to hide the truth about the epidemic. At the port of call in Livorno, Italy, the doctors did nothing to detain the ship.

Captain Jean-Baptiste Chataud himself was eager to deliver his cargo (bundles of fabric worth 100,000 crowns) before the Beaucaire Fair. He moored his boat at Le Brusc, near Marseille, and discreetly notified the ship owners. They made use of their relations. They appealed to the aldermen of Marseilles to avoid a brutal quarantine which would consist of isolating the ship (and its cargo) in the open sea for 40 days. All of them considered the plague to be a thing of the past and detached themselves from the matter.

Finally, they asked the captain to go back to Livorno to look for a "*net patent*", a certificate attesting that all was well on board. The authorities in Livorno, who did not want to bother with the ship, did not have any difficulty in issuing the certificate. Thus, the *Grand-Saint-Antoine* was placed in "*soft*" quarantine: the sailors had disembarked and were locked up in a quarantine station or free clinic, near the island of Pomègues.

But the men, once ashore, no longer intended to take care of their dirty laundry. They made bundles of it and threw it to the washerwomen over the quarantine station fence…

On June 20, on rue Belle-Table, in a miserable part of the city, a 58-year-old washerwoman, Marie Dunplan, died after a few days of agony. She had charcoal on her lips. The doctors were not careful. How would they make the connection with the Black Death of medieval times? On June 28, in the same neighborhood, a 45-year-old tailor, Michel Cresp, also died. Two days later, it was his wife's turn…

Finally, on July 9, two doctors, the Peyronnels, father and son, went to the bedside of a 13 year old child on rue Jean-Galant. And there, right away, they understood: it was the plague! These two excellent doctors warned the authorities. It was necessary to act quickly… On July 22, a big thunderstorm, accompanied by heat and humidity, accelerated the proliferation of the bacillus. Soon, the epidemic caused a thousand deaths per day in the city. The victims of the contagion died in less than two days.

The houses of the victims were bricked up. Corpses were limewashed… The bishop of Marseilles, Henri-François-Xavier de Belsunce de Castelmoron, adviser of the king and eminent figure of the kingdom, stood out thanks to his exceptional devotion. He put the episcopal palace at the service of the medical profession by taking care of the cleaning of the laundry. He himself walked the streets, assisting and helping the sick, in defiance of death, which finally spared him. The Belsunce course and the high school of the same name recall his heroism.

Another character, the knight Nicolas Roze, stands out from the rescuers. This alderman offered freedom to galley slaves in exchange for their assistance. Under his leadership, the convicts and 40 volunteer soldiers covered their faces with cloth masks and removed, then incinerated, the 8,000 corpses rotting in and around Place de la Tourette.

But the decision was made too late to seal off Marseille in early September, and the bacillus was able to spread inland, so it would take another two years of fighting to eradicate the plague from Languedoc and Provence.

The *Grand-Saint-Antoine* was towed to Ile Jarre, in front of the creeks, and burned on September 26, 1720 by order of Regent Philippe d'Orléans (its remains can still be seen). As for Captain Chataud, he was imprisoned on the Ile d'If.

Deficits

Cultural

Infallibility

The plague was certainly a disease that claimed huge numbers of lives during the Middle Ages, but in 1700, it was thought people were safe from a new pandemic. Moreover, several doctors (in Marseille) should have diagnosed the plague seen in the sick. However, being sure that the disease was no longer present, they did not do so. Therefore, there was no particular monitoring of the return of the disease at the time, as it seemed unlikely.

Oversimplification

It was estimated at the time that isolating sailors for 40 days was sufficient to avoid any risk of contagion. Consequently, it was accepted that the sailors should undergo a 'soft' quarantine with very little supervision of their actions (such as having their laundry washed outside, for example).

Navel-gazing

The plague was raging at that time in Asia and the Middle East. These were places regularly frequented by the merchant navy and therefore the latter was particularly exposed to the disease. However, these were distant countries and it was considered difficult for the plague to spread to France, especially since quarantine measures proved sufficient to detect the disease in sailors.

Lack of communication

Numerous cases of communication failures were observed. Firstly, the captain, although aware of the deaths on his vessel, did not warn the competent authorities of the risk of his vessel harboring a disease. Seafarers too could have alerted someone, but it seems that their eagerness to get out of quarantine deterred them from doing so.

The Italians, although they were well aware of the epidemic on board the ship, let her go for the first time and then issued a clean license. There was clearly a lack of communication between all these parties (by "omission" in these cases).

Organizational

Productivism

This deficit is obvious: the greed of the captain, the aldermen, the shipowner and the owner was clear to see. The opportunity to sell the fabrics at the Beaucaire fair removed any notion of precautionary principle.

Dilution of responsibilities

The real responsibilities were very much shared. Concealment by the patients themselves was witnessed at the beginning, but the ship's doctor and the captain should have declared the suspicious deaths between Tripoli and Livorno (about 10). The shipowners who discreetly came to consult the captain and the state of the goods at Le Brusc should have enforced the quarantine rules. The shipowner and the Marseille aldermen turned a blind eye because of the fair in Beaucaire; to cover themselves, they even asked the captain to return to Livorno to obtain a certificate (of convenience?).

Managerial

Lack of feedback

The plague raged for several years in France during the 15th and 16th centuries. In 1680, Brittany had even decreed a ban on ships coming from Spain suspected of being contaminated by the plague from entering the ports of the province. The last epidemic across all of France dates back to 1670 and had seemingly been forgotten by 1720.

No cindynic procedures

Some procedures such as those of Brittany, Bruges were known. In Marseilles, they opted for a soft quarantine… The exact causes were unknown; it was difficult to defend oneself. Isolation was one of the most effective means if it was strictly enforced.

The theory of contagion by microscopic parasites was defended around 1550 by Fracastor.

No cindynic training

Nothing but a few sanitary principles like putting a cloth on your face or burning everything.

No preparation for crisis situations

Despite the existence of quarantine to prevent an infectious outbreak, there was at the time no preparation from the authorities and the population to react in the event of a crisis situation (a proven epidemic). Responsibilities and actions to be taken were not clearly defined. This had the effect of delaying the containment of the city of Marseille, allowing the epidemic to spread throughout Provence and Languedoc. It was therefore necessary to isolate the entire regions in order to contain the infection. The lack of speed and coordination is clear in this case.

A6.3. The 1794 explosion of the Grenelle gunpowder depot

The explosion of the Grenelle gunpowder depot on August 31, 1794 is a remarkable event in the history of industrial risks. Indeed, this catastrophe led to the first laws on classified installations. The imperial decree of October 15, 1810 and the ordinance of January 14, 1815, after feedback, set the first regulations concerning installations, factories and workshops presenting a danger or inconvenience to the environment.

The Château de Grenelle dating from the 17th century was bought by the French state in 1751 for the military school. The convention installed a gunpowder depot there. Chaptal, the famous chemistry teacher, was appointed director of the manufacture of Grenelle powders by Lazare Carnot.

The new working methods he recommended increased the daily production of saltpeter from 8000 to 35,000 pounds (~3.7–16.3 t). Jean-Antoine Chaptal never ceased to explain to the *Comité de Salut Public* (French Public Health Committee) that this rapid progression could not be achieved without major risks. The 2,000 workers who piled into the workshop were making black powder in an overheated atmosphere without any protective measures having been taken until then.

On August 31, 1794, at 7:15 am, 65,000 pounds (30 tons) of powder exploded in the depot of the Château de Grenelle located near the military school. Trees were uprooted and buildings knocked down. The day before, 50,000 pounds (~23.3 t) of powder were sent to the borders. Two days earlier, 100,000 (~46.5 t) had left the powder store. Luckily, for three months, there had not been so little powder stored in the gunpowder depot as at the time of the explosion. The noise generated by the explosion could be heard all the way to Fontainebleau. Rescuers recovered the injured and pulled hundreds of victims from the rubble. The Charité and Gros-Caillou hospitals were overwhelmed by the massive influx of patients. The police administration was responsible for keeping a special watch on public institutions. The committee for public salvation and general safety, which was responsible for monitoring the arsenal and the detention centers, secured in Meudon the stocks of gunpowder, saltpeter and sulfur that had escaped the explosion. The explosion caused significant material damage. Its force was such that the houses in the neighborhood were shaken, bridges were cracked and the seals on the doors of the Observatory's underground passages were broken. This catastrophe resulted in the deaths of more than 1,000 people.

While these major disasters are quite rare, accidents and explosions destroying all or part of a mill or gunpowder depot were unfortunately very frequent.

Archives have left us the reports of Pelletier, who, as a young artilleryman, was in charge of an inspection mission. For the year 1705 alone, he recorded 11 accidents involving 9 different mills, most of which were naturally due to fires. On April 24, 1705, 150 pounds of powder was destroyed at the Maromme mill and, 15 days later, a second fire destroyed another 140 pounds.

In this case, the losses had been minor, but they could have been greater, as in Colmar on January 24, when a fire ravaged the granary, destroying 11,668 pounds of powder. Four months later, a storm damaged the mill, again resulting in the loss of 1,050 pounds. The other threat to the mills was flooding due to their location on the banks of a river. This was the case in Fenestrelles in 1705 and in Toulouse in 1712. Inefficient "logistics" (storage and transport) caused the loss of large quantities of powder. Here again, inspection reports mention a very high number of incidents leading to wastage of the precious powder.

Powder depots were too few in number, too small and poorly designed: barrels were stacked five or six high (or even more) and those in the lower row burst under the load. The powder was stored for too long, sometimes in damp stores, and turned out to be "spoiled" when used. The barrels were of poor quality and lost their load on the road, to the great commotion of the populations of the cities they passed through.

Of course, the role of the inspectors was to point out what was wrong, and the long list of incidents should not hide the essential point: during the four to five centuries of the *Ancien Régime*[2] mentioned here (say from 1350 to 1789), the King's armies and wealthy individuals had black powder of increasing quality. On the eve of the Revolution, manufacturing methods reached a degree of sophistication and efficiency that would do credit to European charcoal burners, saltpeter makers and powder manufacturers, if it were not for war purposes.

2 The political and social system of France from the Late Middle Ages until the French Revolution of 1789.

Deficits

Cultural

Infallibility

This deficit cannot be retained: indeed, the danger was recognized by all.

Oversimplification

The making of the powder, although dangerous, was considered simple: mixing charcoal, sulfur and saltpeter in the proportions of 1, 1, 6. The problem was that the preparation of the three ingredients and their mixing often took place in the same enclosed space.

Navel-gazing

Numerous accidents and explosions were recorded throughout the 18th century, but no safety measures were put in place in Grenelle.

Lack of communication

N/A. The exact circumstances of the accident are still unknown.

Organizational

Productivism

Chaptal recommended new production methods that increased saltpeter production from 3.7 tons/day to 16.3. 2,000 workers manufactured powder in a superheated atmosphere. The needs of the revolution for the war against the other European countries – Holland, Prussia, England – were enormous: 1 million men fighting after the mass rising of 1793. Failure to meet production targets was considered treason punishable by the guillotine.

Dilution of responsibilities

Chaptal was investigated and then cleared and despite his fame, he could have been guillotined like Lavoisier. It was the soldiers of the revolution who were pushing for production despite the reported risks.

Managerial

Lack of feedback

In the 13th century, Marcus Graecus, who developed a mixture very close to the one seen in the explosion (1, 1, 6), recommended that all handling be carried out only outside the premises.

Numerous catastrophic explosions are known: the city of Lubech in 1360, Boston in 1645, the Verdun mill in 1727 which caused hundreds of deaths and destroyed a large number of the towns concerned. But the safety conditions were not really improved.

No cindynic procedures

The existence of firemen was very recent (1722). The battalion of the Paris fire department was created only in 1811. The work was known to be dangerous but no rules of protection, prevention and intervention had been put in place. The improvised relief efforts, despite their good will, were very quickly overwhelmed.

No cindynic training

N/A.

No preparation for crisis situations

The clearing of rubble to save hundreds of injured people was totally improvised with rudimentary means. The Charité and Gros Cailloux hospitals, which were overwhelmed, had never been given the task of preparing for such an accident.

A6.4. The sinking of the Titanic April 15, 1912

If there is one catastrophe whose analysis could best illustrate the cindynics approach, it is this one.

Indeed, as we will see, the study of the shipwreck perfectly shows the various deficits that inevitably led to dangerous situations.

Let us briefly recall the facts before proposing a cindynic analysis of the causes.

At that time, White Star Line was one of the most important shipping companies. Founded in the 19th century by Thomas Ismay, this British company had been owned by Americans since the beginning of the 20th century. Bruce Ismay, who remained director of the White Star Line, intended to resist his competitors and win the Battle of the Atlantic.

In 1907, he planned to put into service three super liners: the Olympic, the Titanic and the Gigantic. On July 31, 1908, the project was approved and an agreement was signed with the builder.

As in the past, the construction of the three White Star Line liners was entrusted to Harland and Wolff of Queen Island, near Belfast, Northern Ireland. Founded in 1847, the company, which employed more than 14,000 people and could build eight ships at a time, was the largest shipyard in the world. The plans for the three liners were drawn by its team of engineers, led by Thomas Andrews.

On March 31, 1909, the construction of the Titanic began. On September 18, 1911, the White Star Line announced that the maiden voyage would officially begin on March 20, 1912, but unforeseen circumstances forced it to postpone the departure date.

Indeed, the Olympic, which had been in service since 1910, collided with another boat, the cruiser Hawke, and therefore needed to be repaired. Two months later, she set sail again, while the Titanic was receiving the finishing touches to its construction. The construction of the transatlantic liner was completed on March 31, 1912. The first tests took place at the beginning of April. Once the tests were completed, the Titanic set sail for Southampton to begin its maiden voyage.

With its double riveted steel plate hull and 16 compartments separated by 15 watertight bulkheads, the Titanic offered maximum safety. If one of the compartments was ever hit, the bulkheads could be closed from the bridge by means of electric control. In the event of massive water ingress, the bulkheads could also be closed manually or automatically by means of safety floats.

The transatlantic liner was also equipped with eight pumps with a discharge capacity of 400 tons of water per hour. With such a system, the Titanic was virtually unsinkable. Indeed, two compartments could be flooded without endangering the ship, the others ensuring its buoyancy.

In addition, the Titanic was equipped with a state-of-the-art navigation system for the time.

Its colossal dimensions also protected it from the most violent storms. The liner also had smoke detectors in critical areas. The alarm bell ensured a quick response by employees in case of fire. To complete the picture, an acoustic detection device was used to detect submerged obstacles. No wonder that in 1912 the Titanic was somehow considered a masterpiece of naval architecture!

Other technical innovations made the Titanic a real success. Unlike other liners, it had two conventional 15,000 horsepower steam engines and a 16,000 horsepower low-pressure turbine.

The total power transmittable to the propellers was 46,000 horsepower. Steam was produced by 29 boilers and 159 furnaces.

The electrical energy required to light the premises (10,000 light bulbs) and to operate the various appliances (for cooking, heating, cooling and ventilation) was provided by four 400-kilowatt dynamos.

The Titanic also had a very *avant-garde* telephone system for the time. It was a double installation consisting of a central office for navigation and an internal system. The first connected the command bridge, the foredeck, the crow's nest, the engine room and the rear compartment. The second, with a capacity of 50 lines, allowed the guests in the luxury cabins to communicate with the various services (bar, restaurant, etc.). The Titanic also had a high-powered Marconi-type wireless telephone system, which allowed the transmission and reception of telegraphic messages.

The Titanic set sail on April 10, 1912 from Southampton in the United Kingdom to Cherbourg in France and to Queenstown along the way to pick up passengers. She left Queenstown on April 11, 1912 before setting sail one final time.

On April 15, 1912 at 2:00 am, the Titanic sank south of Newfoundland in the North Atlantic after striking an iceberg. The number of missing is estimated at 1,500 and the number of survivors at 700.

The technical development of ships reflected the fierce competition between shipping companies: Cunard, White Star, Compagnie Générale Transatlantique, Hansa, Hamburg Amerika, Nord Deutscher, etc.

The Titanic seemed to offer an unprecedented level of security. White Star President Bruce Ismay had decided to take part in this inaugural crossing.

Captain Smith, 62 years old, was the most experienced commander sailing the Atlantic at that time. The ship's design engineer, Thomas Andrews, was also on board.

Deficits

Cultural

Infallibility

The carefully orchestrated press campaign talked about masterpieces of naval architecture and unsinkable boats. Some journalists nicknamed it "the unsinkable".

Captain Smith himself said of shipwrecks, "With such a ship, it's a problem that no longer exists!".

At 8 am on April 15 in New York City, as the tragedy unfolded, Philip Franklin, Vice President of the White Star, said, "The Titanic is not in danger of sinking. The ship is unsinkable and the passengers will only experience a few inconveniences!".

Despite the alarming news, the company continued: "In any case, the boat cannot sink, there is no risk for the passengers. We are convinced that the watertight bulkheads will prevent it from sinking".

Oversimplification

The number of lifeboats was regulatory, even higher than requirements, but the rule dated from 1894 and concerned the units higher than 10,000 tons for 50,000 tons for the Titanic.

In any case, being considered unsinkable, it was not necessary to provide more lifeboats that would never be used.

Navel-gazing

Captain Smith, the previous year aboard the Olympic, the sistership, could not avoid a collision in the Solent off the Isle of Wight with the Hawke, a light cruiser. This showed a maneuverability problem, but nothing was changed for the Titanic.

The captain maintained his route despite several telegrams indicating the presence of ice: by the Coronia, by the Beltic at 1 pm, by the Greek steamer Athenai and by the Mesaba at 9:40 pm. The captain decided to keep the planned route from the North and to maintain his speed while Mount Temple informed him of the presence of icebergs and diverted him towards the South. Smith probably wanted to demonstrate the superiority of the Titanic and his own at the same time.

Lack of communication

The Titanic had a modern wireless installation but not the other boats including the Carpathia.

The operator did not transmit all iceberg alert messages to the captain, especially the last ones. He preferred to take care of the telegrams received and to be sent by the rich passengers of first class. Indeed on board was: the billionaire John Astor, Major Butt aide-de-camp of the American president, Charlotte Cardeza, a billionaire from Philadelphia, Benjamin Guggenheim, the king of copper, Charles Hays, the king of the Canadian railways, Isidor Straus, a big store owner in New York, and Widner, the king of the tramways, one of the biggest fortunes.

The lookouts had no binoculars, forgotten in Southampton, and did not see the icebergs until it was too late.

The procedures for international distress signals had just changed. They went from the old CQD to the more audible SOS.

The two messages sent by the Titanic increased misunderstanding.

The scope of the wireless technology was limited.

Organizational

Productivism

The Titanic was turning too slowly. This problem was reported during the short tests carried out before the crossing. The dimensions of the rudder and the size of the propellers should have been increased. But the maiden voyage was already scheduled and nothing was corrected.

In order to meet the deadlines, the Irish yard used the same steels as for conventional ships; we see that this may have been of great importance.

The tests lasted only one day on April 2, 1912!

To prove the speed of the Titanic and at the request of many passengers, the objective was to arrive before evening in New York. Indeed, to take into account the unloading, the disembarkation of the passengers and the formalities, it was better to arrive in the afternoon. Smith therefore decided to keep going at a fast speed and to take the fastest orthodromic route, known as the Great Circle.

Dilution of responsibilities

The search for the causes of the accident shows several simultaneous failures: the lookouts, the captain, the radio operator, the designer, etc.

Questions have been raised about the impact of having the company president, the design engineer, several busy billionaires on board, and so on.

Was the captain the only master on board free to make his own decisions?

Managers

Lack of feedback

Maneuverability tests were not studied carefully enough, during tests that were too short.

No cindynic procedures

The large number of casualties, especially in third class, can be attributed to the lack of emergency plans, evacuation procedures and alarms. Cabin attendants went to wake up passengers in their cabins!

Captain Smith hesitated at the last moment between having the passengers boarded on the promenade deck or on the upper deck.

No cindynic training

The crew was unfamiliar with the handling of the lifeboats, some of them overturned. A fire and rescue drill should have been held on Sunday, as required by White Star regulations. But it did not take place.

No preparation for crisis situations

The company had no reliable information until almost 24 hours after the tragedy. The officers were overwhelmed. The lifeboats did not have the necessary survival elements, such as drinking water. The seamen did not know how to handle the lifeboats.

A6.5. October 29, 1929, the Wall Street Crash on Black Tuesday

World War I ended in 1918 thanks to logistical and military support from the United States. While the world economy of the 19th century had been led by all-powerful England, the 20th century began with American hegemony. Victory and strong economic growth (more than 15%) developed American confidence. It is a period of wealth and optimism: the Roaring Twenties. A decade of pleasure and hope, despite prohibition, consumer goods became cheaper. Technological innovations and productivity improvements seemed to prove that anything was possible. Low interest rates allowed Americans to borrow and go into debt without counting to consume. In 1927, Lindbergh crossed the Atlantic and America was the strongest. The stock market offered for this new era a certain prestige, thrills and the promise of becoming rich. Major companies such as General Motors, Dupont de Nemours and RCA (Radio Corporation of America) saw the value of their shares multiplied by 10! The Dow Jones index rose from 100 to 400 between 1923 and 1929. Americans borrowed to play on the stock market despite the increase in rates by the government to calm this frenzy.

The economist Babson, one of the few to do so, warned of the danger, the recession was already noticeable. The first warnings began on Saturday, October 19, when the market showed a significant decline, the first in years! On Thursday, October 24, Black Thursday, huge quantities of shares were traded, values began to fall; General Motors lost 80%! The vice-president of the Stock Exchange, thanks to a common fund of bankers, managed to stop the fall. President Hoover, himself, intervened to reassure Americans about the health of their economy.

Tuesday, October 29, 1929 was the worst day of the Crash. As soon as it opened, the biggest speculators gave in to panic and sold millions of shares in a few hours, sometimes at a tenth of their value. The shares of RCA (telecommunications: a flourishing sector), which were quoted at $420, were sold at $26. Demented cries resounded throughout the stock market: “I’m ruined, ruined”. Brokers were crying, the reputation of the place and the bankers was destroyed. At closing, 16 million shares had been sold off. Several speculators, such as the president of the Rochester Gas and Electric Corporation, committed suicide.

This Black Tuesday remained in all memories, but the prices continued to collapse for several months after the crash.

The victims and in particular the millions of small savers sought those responsible. Hoover was severely criticized and lost the election against Roosevelt. Famous economics professors were discredited by their optimistic announcements, blind to the impending disaster. A Great Depression hit the United States with 14 million unemployed in 1932. Hoover refused to grant state aid to the poor, which could have alleviated the suffering and revived the ruined economy. Thousands of businesses, especially agricultural ones, went bankrupt.

Roosevelt was elected and implemented a major national recovery plan with the “New Deal” as well as protectionism. As a repercussion, Europe, which had to pay back loans in the United States, went into recession and witnessed extreme behavior.

Adolf Hitler became chancellor in 1933, his party stigmatizing the depression and misfortune coming from elsewhere.

Deficits

Cultural

Infallibility

Hoover displayed rigidity; they said he knew everything about everything and was always right. He said: “Lack of confidence in the American economy is foolish. The crisis will end in less than 60 days. Fortune is just around the corner”. The Empire State Building project was initiated by Raskob, who had won a huge fortune on the stock market. On this model, many small shareholders had mortgaged their houses and spent all their savings to play the stock market, all sure to become rich. On October 15, Mitchell, a stock market specialist, said: “The industrial situation in the United States is excellent”. Fisher, an economics professor at Yale, said, “The stock price has reached a high level and should stay there for a long time!”

Oversimplification

During his victorious campaign in 1928, President Hoover said, “Today we are closer than ever to the abolition of poverty and despair!”.

Navel-gazing

No one listened to Babson’s pessimistic predictions. If the others lost, everyone was sure to make a fortune.

Lack of communication

Despite the widespread use of teleprinters throughout the country, the price evolution was known more than an hour late, up to five hours after closing on Black Tuesday.

There was no consultation between the Stock Exchange, the banks and the state.

Organizational

Productivism

The stock market crash should not have hidden the crisis that was brewing – too much production, too much credit and too many dreams.

Dilution of responsibilities

On October 28, the bankers stated in a press release that they were not responsible for stabilizing prices. It was the stock exchange that was responsible for its operation. The following were blamed: Hoover, brokers, stockbrokers and the American economic model. Coolidge before Hoover felt that it was the state of New York that controlled the actions of its stock exchange and not the federal government.

Managers

Lack of feedback

Past events could have served as a lesson. In the 17th century, an irrational passion for Dutch tulips had turned them into a product of huge speculation. The fashion quickly passed and the prices collapsed.

In 1720, in England, the collapse of the South Sea Company on the stock exchange caused a long economic depression. In 1873, in the United States, a speculative surge on the Northern Pacific Railroad caused financial panic. On September 3, 1929, the economist Babson, like Cassandra, warned of the possibility of a crash. On October 24, the first fall in prices should have alerted all professionals, but speculation continued.

No cindynic procedures

At the time, there was no procedure to prevent and react to a massive fall in prices. Regulatory bodies were set up after the disaster.

No cindynic training

N/A.

No preparation for crisis situations

The dizzying drop in prices surprised everyone, shareholders, professionals and even academic specialists such as one of the most eminent economics professors, Irving Fisher.

Appendix 7

Bhopal Study

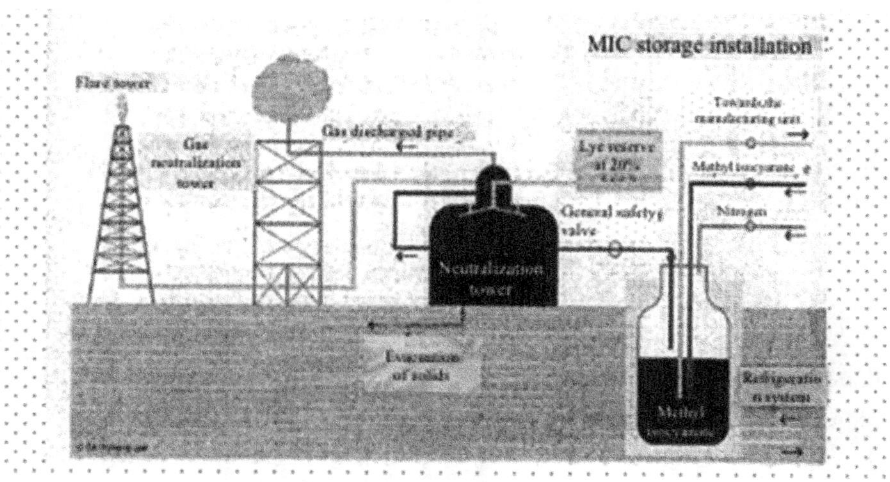

Figure A7.1. *Diagram of the MIC storage facility*

This excerpt is taken and translated from the journal *Préventique* (August–September 1986).

This figure does not provide explanations but shows interactions between about 20 descriptive factors.

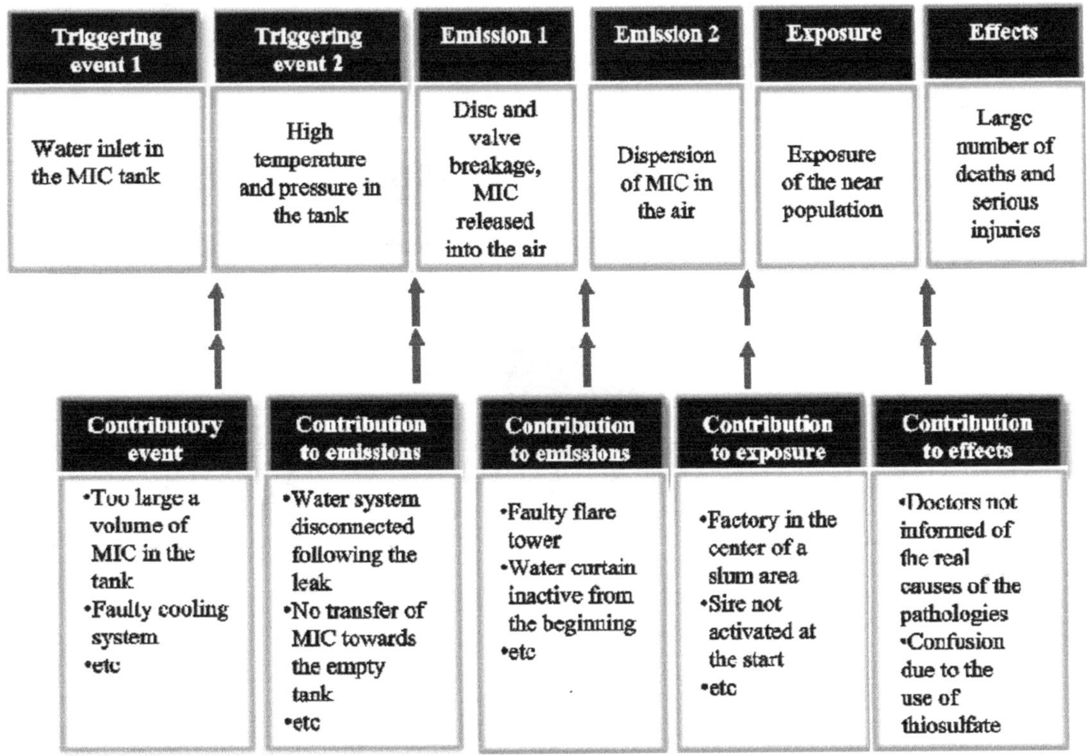

1a – An explicative model of the causes of the Bhopal accident according to J. and R. Kasperson

Table A7.1. *Causal and linear analysis by J. and R. Kasperson*

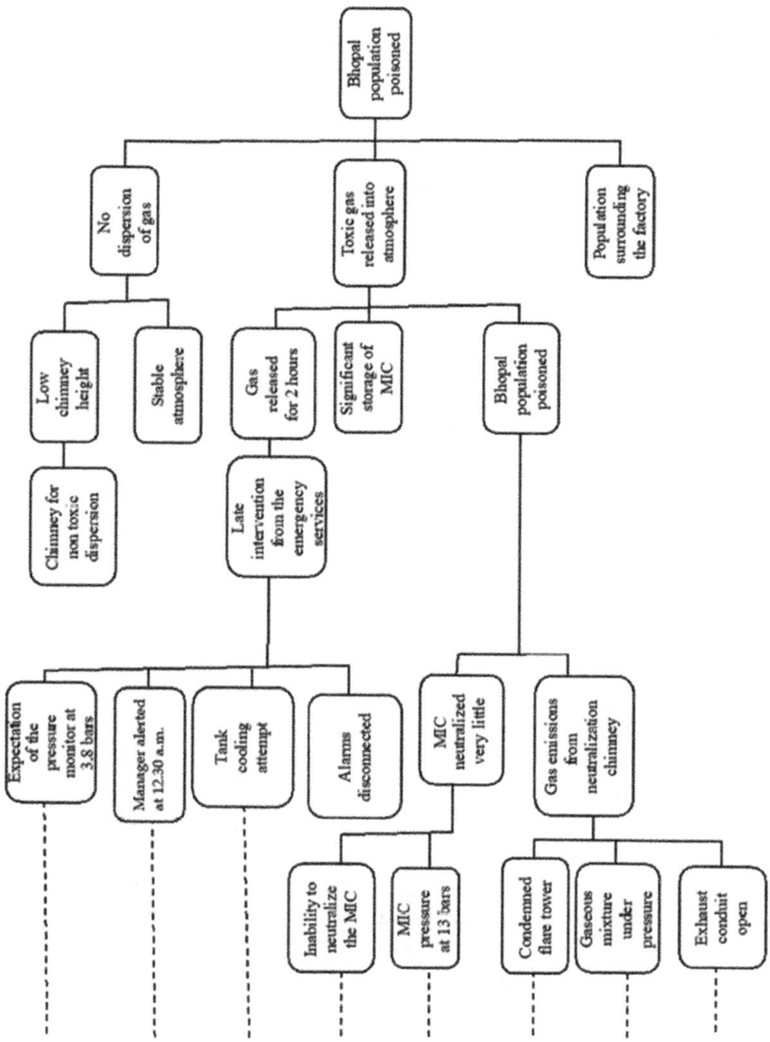

Figure A7.2. *Excerpt from the multi-causal tree approach to causes according to Grenouillet* et al.

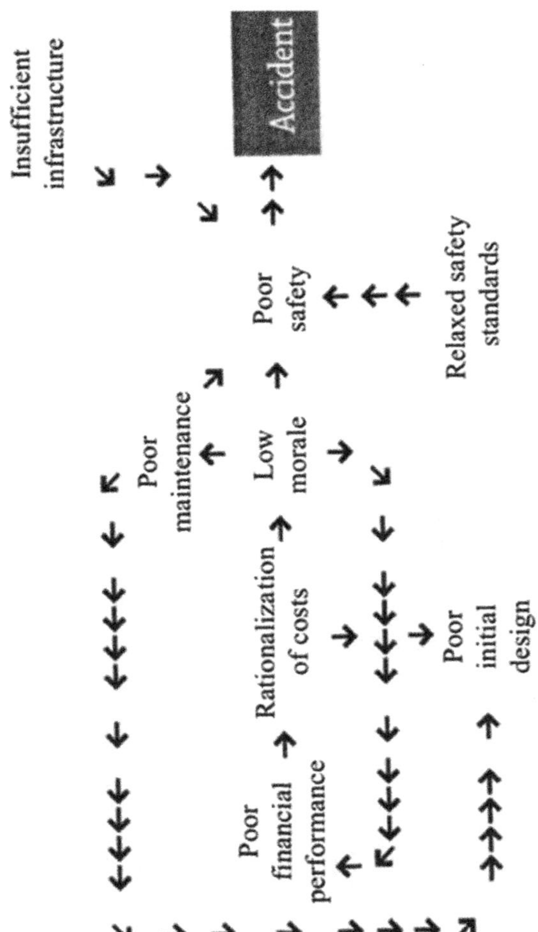

Figure A7.3. *Systemic approach according to Pauchant and Mitroff*

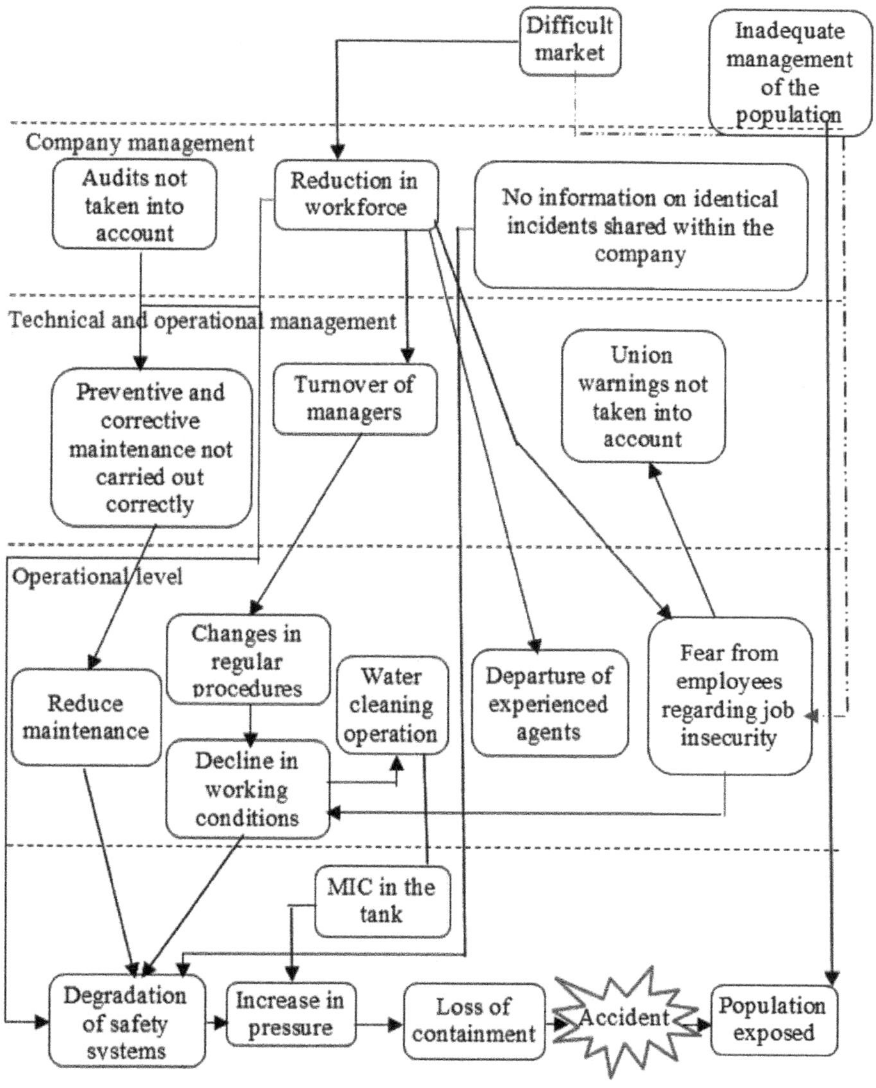

Figure A7.4. *Filter approach according to Le Coze* et al.

Appendix 8

More Information About Bhopal

Union Carbide (UC), one of the most significant multinational chemical companies in the US, decided in the 1950s to produce pesticides with the slogan: "holding the line for a richer harvest". In 1957, three Carbide researchers developed the formulation of Sevin, a pesticide; using MIC, Sevin was presented as a "miracle" pesticide, acting on almost all parasites. Its advertisement praised its non-toxicity by presenting a human tasting grapes treated with Sevin. Carbide was not completely transparent about MIC. While the company had implemented numerous safety procedures and recognized its lethal nature in case of inhalation, it did not disclose a study that mentioned that MIC produces cyanidric acid under the effect of heat and that an injection of sodium thiosulfate is an effective antidote.

Carbide built an MIC plant in the US in the early 1960s in a highly dangerous chemical complex and presented the plant as a model of safety. In the Carbide plant in the US, the motto displayed was: "safety first". Between January 1, 1980 and the end of 1984, the Federal Environmental Protection Agency (USA) found 67 MIC leaks in the US Carbide plant.

In the 1950s, India planned a "green revolution" aimed at achieving food self-sufficiency. This project included significant development of agriculture, both extensive and intensive. From 1966 to 1971, the Green Revolution in India resulted in a threefold increase in fertilizer consumption. The need for pesticides exploded, and local production was not enough.

In 1966, India decided to use foreign manufacturers. The Indian subsidiary of UC was well known and recognized in India (UCIL, Union Carbide, India Limited), manufacturing batteries, plastics, photo films and a fleet of trawlers in 14 factories.

The Sevin representative for Carbide was Eduardo Munoz. In 1966, he obtained the import of 1,200 tons of Sevin. In exchange, Carbide set out to build a factory in India over five years that would manufacture Sevin.

The plant was finally built four years late. The manufacturing of the MIC did not start until 1980.

The site chosen for the construction of the plant was Bhopal, capital of Madhya Pradesh in central India, because of its geographical location, railway station, electricity production and the presence of sand (necessary for the "formulation" of Sevin). The site was along the railway line, 1,500 m from the station, close to the city center and adjacent to a slum. Between 1961 and 1980, the population of Bhopal rose from 225,000 to 700,000 inhabitants.

In India, "Carbide offered Sevin as a solution to Indian farmers' troubles, as their crops were then being regularly ravaged by insects"[1] and promised earnings of 5 rupees for the purchase of 1 rupee of Sevin.

At the end of the 1960s, UC obtained authorization to produce 5,000 tons of Sevin per year; Eduardo Munoz estimated the need at 2,000 tons.

In 1960, UC used the Hay rating system, which favored large production units, and decided to devote $20 million to the construction of the Bhopal plant.

The Bhopal plant was built on the same model as the one in the US and had to operate continuously, implying significant storage of MIC. This was in line with the US culture.

A survey by Munoz in Europe informed him that in Germany, MIC was only manufactured as and when needed. For German (and French) engineers, storing MIC was akin to placing an atomic bomb in the heart of the factory.

According to the UC project manager, "the Bhopal factory will be as harmless as a chocolate factory".

The government of Madhya Pradesh prohibits the construction of factories releasing toxic products where the prevailing winds risk blowing these products into densely populated areas. UC only mentioned Sevin (non-toxic) in their record and not the components used in its manufacture.

A Bhopal municipal official who opposed the construction of the plant, citing the law, was transferred after UCIL donated $2,500 to the city of Bhopal for the creation

1 https://www.pophistorydig.com/topics/union-carbide-1950s-1980s/.

of a park. A member of the ruling party became UCIL's lawyer. The former head of the local police force was made responsible for the plant's security and the nephew of the minister of education was appointed director of public relations.

The Bhopal factory did not have the same level of safety as the factory in the US. In fact, sales forecasts quickly turned out to be exaggerated due, in particular, to the appearance of cheaper competitors. Moreover, as the Indian government limits the participation of foreign companies in their Indian subsidiaries, UCIL was demanding cuts in the factory's budget.

The first director of the Bhopal plant was named Warren Boomer. He was an American engineer and a specialist in hazardous products. He was replaced in 1983 by an Indian financial director of the subsidiary of a multinational company in India.

Cultural difference between Muslim and Hindu engineers led to problems. According to the plant manager: "the Muslim exposes the problem while the Hindu procrastinates before blaming bad fate [...] The Sikhs, with their beards, have difficulties putting on a breathing mask [...]".

A few months after the start-up of the plant, water pollution led to the death of several cows. The plant immediately compensated the owners and had the water tested, but did not disclose the results.

Several thousands of people lived in a slum in the immediate vicinity of the Bhopal plant.

In 1981 and 1982, a severe drought led to numerous farm closures, impoverishment of farmers and a significant drop in pesticide sales, especially Sevin.

The medical staff at the Carbide plant in Bhopal had not received any information on the pathology of gas accidents that may be caused by methyl isocyanate.

The plant was equipped with an alarm siren, but the information broadcast, including the evacuation order, was intended only for plant personnel.

The government of Madhya Pradesh refused to validate a comprehensive evacuation plan proposed by Warren Boomer in the run-up to the elections.

Carbide provided a luxurious suite at the permanent disposal of the Prime Minister of Madhya Pradesh, in his research center in Bhopal.

On December 23, 1981, following a routine maintenance operation, a technician was exposed to droplets of phosgene. Not following the decontamination instructions to the letter, he removed his mask too soon and died a few hours later. On February 12, 1982, 25 workers were intoxicated (no deaths). The plant's management stated that no leak resulting from failures ever exceeded the lethal toxicity level (this level was not listed in any manual). On October 5, 1982, an MIC leak led to the evacuation of the plant. Fortunately, the wind was blowing towards the countryside. In May 1982, a safety audit carried out by UC in Bhopal revealed more than a hundred anomalies (including recurrent MIC and phosgene leaks and a very significant turnover of the factory workers, as well as a lack of staff training). Many workers and technicians spoke only Hindi (due in part to the turnover), while the instruction manuals were written in English.

In 1982, Carbide sold 2,300 t of Sevin in India, less than half of the planned capacity of the Bhopal plant. Forecasts for 1983 were even lower. In 1983, UC's management imposed a savings plan in Bhopal. The teams, including the one assigned to the manufacture of MIC, were cut in half. 150 workers were pooled to intervene as needed. Maintenance was reduced to a minimum. In the fall of 1983, the factory idled according to needs (but more than 60 t of MIC was stored). The manager shut down the main safety systems, including the MIC refrigeration and flaring, considering the plant almost at a standstill.

In September 1984, a safety audit at the Carbide plant in the US showed numerous anomalies in the MIC manufacturing and storage system and concluded that there was a potential risk of a "serious incident". The report was circulated to 16 people but the Bhopal plant manager was not a recipient.

During the summer of 1984, UC management secretly decided to dismantle the Bhopal factory (the sale of Sevin in India for that year did not exceed 1,000 t, i.e. less than a fifth of the nominal production) and transfer the production units to other third world countries.

On December 2, 1984, tanks 610, 611 and 619 contained 63 t of MIC (beyond the authorized quantity, allowing the injection of solvent in the event of a chain reaction), the refrigeration circuit was stopped and the flare was extinguished.

At 8:00 p.m., the technicians proceeded with a pressure cleaning of the pipes, forgetting to place a plug on the pipes going to the MIC tanks (the person in charge did not know the maintenance rules, having newly arrived). At 11:00 p.m., a major MIC leak was noted, but the shift manager did not believe in the accident and at midnight he ordered a "tea break". At midnight, tank 610 exploded, the order to evacuate the plant was given and the wind blew towards the slums and the city center.

On December 3, around 12:30 a.m., the doctors on duty at the hospital saw an influx of patients. Many died on arrival. They learned of the accident at the Carbide factory but did not identify the dangerous substance. The factory doctor, contacted by phone, told them that methyl isocyanate was not a deadly gas.

Very few victims were compensated. UC paid several million dollars, but no convictions were made: there was a jurisdictional problem, and UC was sold to another company.

Appendix 9

Collection of Information on the Queen Mary II Gangway Accident

A9.1. The facts

It was approximately 2:15 p.m. on Saturday, November 15, 2003, when a gangway linking the quay to the Queen Mary II (under construction and scheduled for delivery on December 22, 2003) collapsed, causing 46 people to fall 18 m, 16 of whom died and 30 were injured, 18 of whom were unable to work for more than three months.

Among the elements of the context, the gangways were also referred to as "cut" although the latter technically corresponds to an opening in the hull of the vessel or under the "planks". They were also referred to under the torque number of the vessel at which they were installed. Thus, the damaged gangway was designated by the number C292/1 (see Figure A9.3).

The 46 people on the gangway came from two groups of visitors accompanied by executives from SA Chantiers de l'Atlantique and a group of employees from the cleaning company who went on board, under the direction of a Chantiers manager, to dust off electrical cabinets.

A tradition allowed managers and supervisors and, under certain conditions, those of subcontracting companies, to show ships under construction to relatives in the evenings or at weekends.

Excerpts taken and translated from [NIC 12].

A large number of visitors were expected during this afternoon of Saturday, November 15: a total of 1019, with 796 between 2:00 p.m. and 3:30 p.m.

Two checks were carried out at two different points by the accompanying manager. Finally, a security guard from the subcontracting company International Security Province (ISP), posted at the access gate, had just been instructed in the morning not to write down the names of the visitors and the accompanying person in order to ensure the smooth flow of the visits.

On November 15, the other access gangway to the ship was unavailable due to painting work. Only gangway C292/1 was available.

Around 2:15 p.m., a supervisor from the construction site went on the gangway accompanied by a group of 11 people. The gangway was empty, but a small group was waiting for its supervisor in the double door entrance, preventing access to the ship.

An engineer in turn walked the gangway with a group of nine people and got stuck.

Then, 26 people (subcontractors) led by another engineer climbed onto the gangway to dust off the electrical cabinets.

Before the 26 people traveling in single file reached the access to the ship, a creaking sound occurred, as the gangway seemed to collapse in the middle. Then, the gangway unhooked on the quay side, then tipped over on the ship side and fell to the bottom of the hold 18 m below.

The *plan rouge* (emergency response plan) was immediately put in place and the injured were evacuated to hospitals in the region.

A9.2. Context

The Queen Mary II had just returned from a second series of sea trials and was being dry-docked to fix a thruster problem.

A9.3. The companies concerned

A9.3.1. *Chantiers de l'Atlantique (CAT)*

Created in 1861 by *Compagnie Générale Transatlantique*, CAT had been a subsidiary of Alstom since 1984. In 2006, this group sold the majority of its marine branch to the Scandinavian group Axe Yards, which took over the Saint-Nazaire and Lorient sites. The shipyards survived this sale and were responsible for monitoring vessels under construction and guaranteeing delivered vessels. At the end of 1997, the shipyards launched a cost reduction plan (Cap 21), followed by a second plan, Cap 21+, to further reduce costs while improving quality and innovation. To achieve these results, the shipyards focused on their "core business" (naval architecture, boilermaking, assembly) by giving an increasing share to subcontracting. This was the case for the scaffolding arm, subcontracted to the company ENDEL for the design, manufacture and installation and for security, subcontracted to the company ISP. Following a call for tenders for the construction of the Queen Mary II, CAT placed an order in April 2002 for external scaffolding for the upper part of the vessel for approximately €2.2 million excluding VAT.

This change had led to a new work set-up. An onboard work leader supervised all onboard work as well as the Methods aspects. On the operational side, each vessel was built under the responsibility of a works coordinator with delegated technical safety and the financial means to carry out his task. Each branch of activity was coordinated by a works manager on board (called a *responsable travaux à bord*, RTB), with authority over:

– supervisors supervising employees for activities relating to the "core business";

– supervisors for subcontracted activities. These supervisors were responsible for helping the subcontractor adapt CAT's business.

As far as safety is concerned, in addition to the work coordinator, there was a Health, Safety and Environment (HSE) department (see Figure A9.1).

A9.3.2. *ENDEL*

ENDEL is the French subsidiary of Suez specializing in industrial maintenance and related services. It was created from the merger in 2002 of Entrepose and Delaitre-Levivier. It is the market leader in France, with 140 locations and 6,000 employees.

At the time of the accident, the contracts with Chantiers de l'Atlantique represented between 70 and 80% of the activity of ENDEL's Saint-Nazaire agency. The Saint-Nazaire agency has, under the authority of a director, a project manager for commercial relations and project validation, a design office for the design of structures and scaffolding placed under the authority of yard managers, whose task is to build scaffolding for the various vessels under construction.

The ENDEL company is specialized in the manufacture of gangways for industrial use only. It is not its job to build gangways for public use. During the hearing, the site foreman (see Figure A9.2) declared that he had a vocational qualification in cooking and then obtained a vocational qualification in metalwork.

A9.4. Contracts between companies

At the end of October 2003, the head of the CAT Logistics Methods Department asked his technician to install a mechanically-welded plank in lot 418 (see Figure A9.3) with a usable width of 1.50 m in order to allow the passage of a 400 kg pallet truck with an 800 kg pallet, a total of 1.2 tons. In a second step, he asked for the modification and the installation of another plank with identical use for lot 413.

NOTE.– in marine construction, a distinction is made between gangways used for the passage of people and those used for the transfer of packages. The latter must be able to circulate traction devices such as wheelie bins.

The CAT company's methods technician deduced that the second plank of lot 413 needed the same length as lot 418, about 12 m. He therefore sent the head of the ENDEL's design office a request for a quote for the construction and installation of two planks supporting a distributed load of 250 N/m^2 and a maximum point load of 2 tons for access C 64 (rear) and for access C292 (front).

On the same day, the ENDEL office manager sent back a quotation for structures, one of 14 m in length for lot 413, the second of 13.5 m in length for lot 418, both validated for a point load of 2 tons and a distributed load of 150 N/m^2, as he considered it impossible, given the present configuration, to foresee a structure for 250 N/m^2.

The CAT methods technician accepted the quotation proposal and his Logistics Methods Manager also accepted it considering that the important point for him was more about the punctual rolling load. ENDEL received a firm order on November 6, 2003.

The two gangways were installed on the quay on November 8, 2003, in accordance with the plans and nomenclature provided by ENDEL's chief engineer and its site manager. The work was completed on November 12, 2003. During the erection, the gangway for lot 413 and the one for lot 418 were interchanged. Furthermore, when the gangway for lot 413 was presented, it was too short since it was only 13.5 m long instead of 14 m as planned. Site superintendent ENDEL had this footbridge removed to extend it by an additional 1.5 m, that is, a length of 15 m instead of the 14 m planned. In addition, both gangways were mounted with a width of 1 m instead of 1.5 m, the scaling of the width of the gangways not appearing on the plans. The people in charge of CAT accepted the modifications because they needed these gangways quickly. However, they decided to have another gangway rebuilt with the right dimensions, while leaving the installation in place until it was replaced.

A new order was placed with the right references. But the assembly on November 14 of the new gangway was done from memory without any plans, taking inspiration from what was done previously. The 1 m wide sheets that made up the floor with wooden plates were not cut to fit the width of the gangway, in accordance with the instructions of the head of ENDEL's design office, but superimposed over the entire length of the plank to save time. In addition, when the gangway was presented in front of the hull opening, bracing to ensure the rigidity of the gangway prevented the passage of the end of the gangway inside the vessel. An operator removed two braces to allow passage. The reception of the new gangway was carried out only visually without a reception report by Chantiers de l'Atlantique. At the end of the assembly of the gangway, a foot of a crane prevented the gangway from being secured to the railings of the quay.

At 2:00 p.m. on Saturday, November 15, 2003, as a result of painting work on other gangways, only the gangway on lot 413 (C292) remained open to provide access to the Queen Mary II for both workers and visitors.

A9.5. Elements of the legal procedure

A9.5.1. *The experts*

Expertise and additional expertise very clearly demonstrated that the gangway had collapsed due to physical instability and not due to mechanical failure.

A9.5.2. *CAT managers*

The head of the HSE department was expected to provide regulatory monitoring for the other departments and monitor compliance with safety rules.

The work coordinator had been delegated responsibility for occupational health and safety at the ship construction site.

The person in charge of the Works on Board Logistics Flow Utility Works had received a sub-delegation concerning safety for the forms of construction and equipping of the vessel. His mission was to plan and coordinate the installation of the accesses and works planned by the works coordinator and to ensure the reception of the various accesses and works.

None of these three people recognized themselves as being concerned by the incriminated gangway.

The Head of Logistics Methods was an engineer specializing in material resistance. However, he did not carry out a check on the quotation which only proposed a load of 150 N/m^2. He did not realize that the plans were not scaled. He did not worry about the conditions of use of the gangway while he was aware of the tolerance granted for the visits to the liner.

The works coordinator did not organize the spacing of the visits to the vessel although he was responsible for it.

During his interrogation, the President of CAT declared that he was unaware of the exact roles of the head of Methods–Strategies of the Shipbuilding department, his functional or operational relations with the works coordinator and the exact organization of the Logistics Methods department. In his opinion, he was in charge of designing, defining and organizing, in the best possible way, all matters related to logistics such as handling, storage or scaffolding. He was also unaware of the degree of competence and intervention of the Logistics Methods Manager.

A9.5.3. *The ENDEL managers*

The person in charge of the design office made it known that, in general, he did not make calculations but chose from the supplier the types of scaffolding that would support the loads specified by the customer.

While it is conceivable to work in this way to design scaffolding that is reduced to a superimposition of modules, such reasoning was deemed unacceptable for the study of a gangway, the calculation of which needed to be done according to its span and therefore the bending moment and shear forces.

In addition, he proceeded by estimation for the realization of the work considering that it could support a load of 150 N/m^2. He did not mention the width of the gangway nor the bracing on the plans.

This head of the design office held an advanced technician's certificate in accounting, then worked as a scaffolder, then as a draftsman and finally as a scaffolding manager.

The site manager did not care about the lack of quotations, which led him to have gangways built 1 m wide instead of 1.5 m wide. He inverted the gangways during assembly, extended one of them and had the gangway rebuilt without plans. He did not notice the absence of bracing, gave instructions not to cut the sheets to the right size for laying on the floor without worrying about overloading and had them overlapped.

The head of the agency testifies that he validated the specifications in the absence of his business engineer and was concerned only with the financial aspect, without taking note of the calculations and plans.

NOTE.–

– the CAT managers and employees involved believed that the accident resulted from a defect in the design and construction of the gangway;

– the officers and employees of ENDEL involved considered that the loss was due to the use of the gangway under conditions that were neither foreseen nor foreseeable.

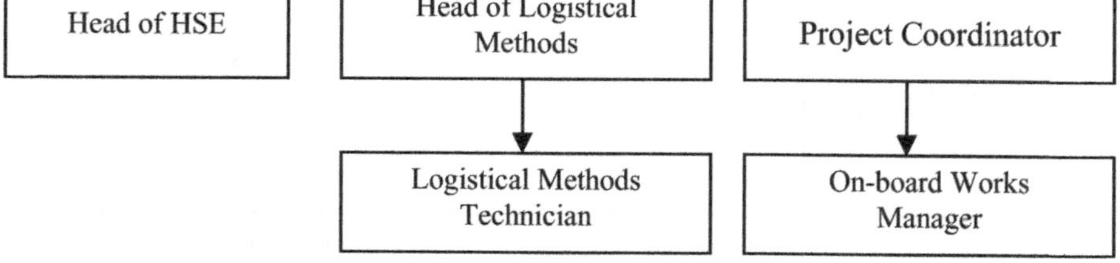

Figure A9.1. *Partial organigram of CAT*

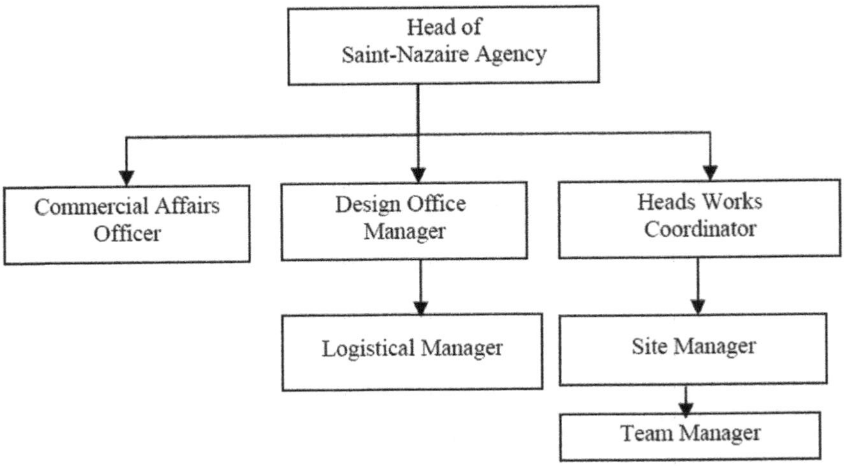

Figure A9.2. *Subcontractor ENDEL's partial organigram*

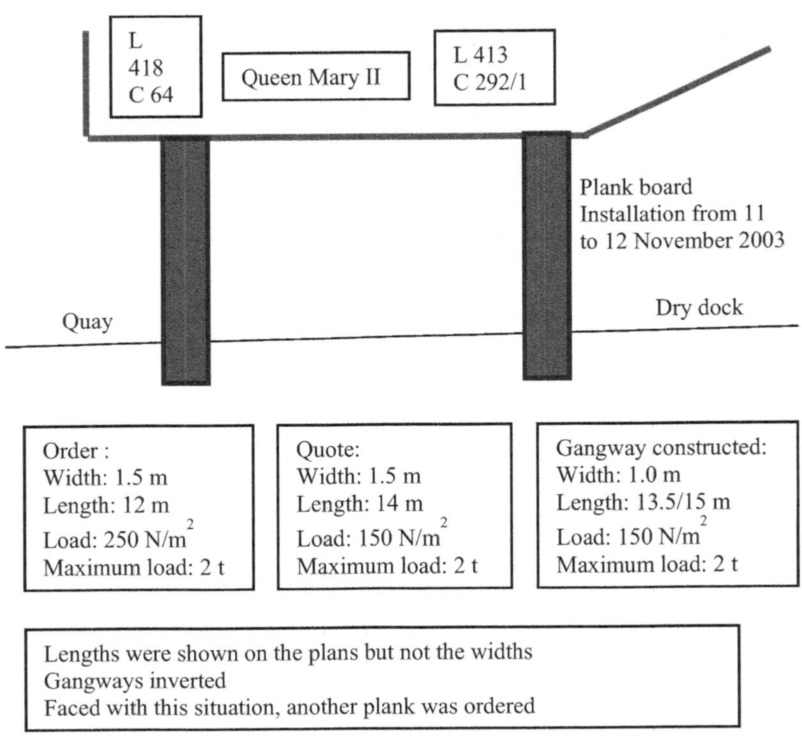

Figure A9.3. *Characteristics and positions of two planks in relation to the ship*

Appendix 10

Queen Mary Accident Cause Tree

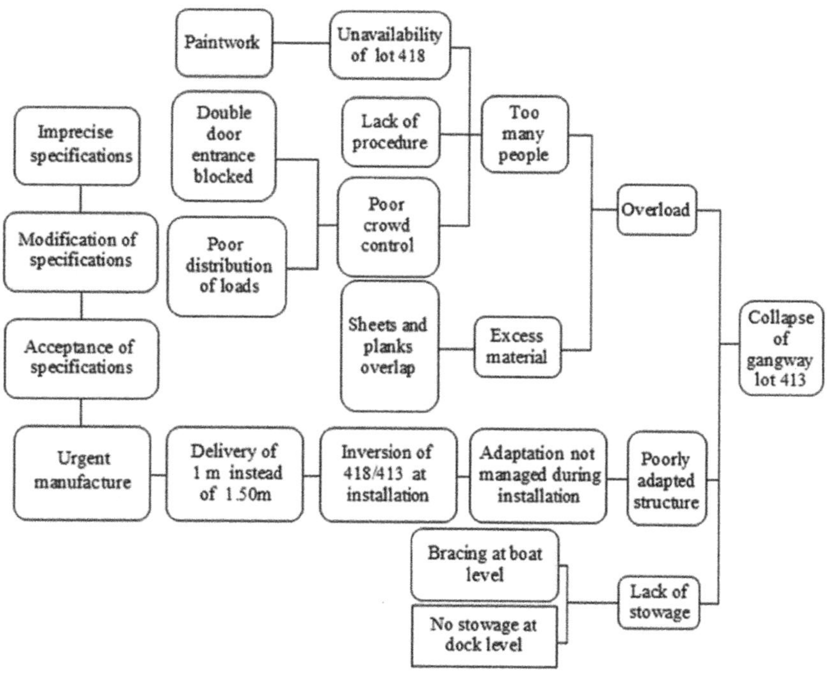

Figure A10.1. *Queen Mary accident cause tree*

Appendix 11

Collection of Information on the Deepwater Horizon Oil Rig Accident

A11.1. Sources of information

– "Final Report on the Investigation of the Macondo Well Blowout" of April 20, 2010. Produced in March 2011 by the Deepwater Horizon Study Group (DHSG) made up of 40 American and foreign experts (academics, public services, consultants) independent of the owner and various operators;

– *Trente ans d'accidents : Le nouveau visage des risques sociologiques* [LEC 16];

– "BP 'Deepwater Horizon' du Golfe du Mexique à l'Afrique : un tournant pour l'industrie pétrolière ?" [MAR 10b];

– *La prise de décision agile : anticiper les risques grâce aux signaux précurseurs* [MAG 17];

– *Deepwater Horizon* (2016). Directed by Peter Berg. [Feature film]. Lionsgate: Vancouver, Canada.

A11.2. Elements identified in the final report on the investigation

– February 2009: in application to MMS, spill associated with blowout deemed unlikely (page 18);

– October 2009: size of the oil and gas field was undetermined (page 20);

– November 2009: hurricane damaging Marianas replaced by Deepwater Horizon (page 21);

– February 2010: a leak noticed on BOP. Failures in pod BOP could not be activated in case of emergency (page 21);

– March 2010: serious well control event – formation fluids flowed in the well bore. The well was side-tracked (page 21);

– March 2010: BP prepared a revised casing design for high pressure (page 21);

– April 4–7: loss circulation occurred solved by pumping high viscosity material (page 29);

– April 14–15: severe loss circulation; logs by Schlumberger showed two oil and gas bearing formations (page 31);

– April 12: Halliburton to use lighter weight nitrogen foamed cement. Tests showed poor compressive strength after 24 hours (page 31);

– April 12: decision to use a long string production casing (page 31);

– April 15: application to MMS for change exploration to production with single long string casing. MMS approved (page 33);

– April 18: Schlumberger arrived to perform cement log (page 34);

– April 20: BP concluded that a cement log was not required (page 38);

– April 20: celebration on board to congratulate operating crew performance for worker safety (page 38).

A11.3. The elements noted in [LEC 16]

– Only seven BP employees among the hundred or so people were present on the platform;

– Halliburton did not sufficiently inform BP and Transocean about the lack of sealing of the cement;

– leakage tests were not carried out correctly; a lot of information existed but was ignored or minimized;

– failure of the BOP (emergency actions did not work) and negligence of the subcontractor who was responsible for its implementation (Cameron);

– the crew did not react to the oil and gas inflow before the effluent reached the platform;

– the fire and gas system did not alert of a potential ignition.

A11.4. The causes of the accident noted by the BP company: elements found in [MAG 17]

– Faulty annular cementing (poorly composed cement, insufficient Halliburton test, checks not in accordance with procedures);

– downhole cement was contaminated by the surrounding physico-chemical conditions;

– poor setting of the cement;

– negative pressure test not correctly conducted and interpreted;

– rise of hydrocarbons in the “riser” not seen because of the high activity on board;

– ineffective and insufficient eruption responses;

– danger zone around the mud tanks was too small;

– BOP failing due to poor maintenance;

– decisions taken were all in the direction of acceleration at the end of the construction site and not caution.

A11.5. The precursor signals noted in [MAG 17]

– On March 8, an eruption of mud and hydrocarbons (which was not taken into account) and an indication of the poor quality of the rock;

– poor dialogue between BP and Halliburton on the results of cement quality analyses;

– the project team decided not to carry out a further assessment of the performance of the cement;

– the use of a spacer (acting as an interface between seawater and sludge) made of recycled sludge, denser than usual. BP chose this solution for money saving reasons;

– the first attempt that failed to bring the pressure down to zero caused the drillers to worry;

– an operator reported a feeling of discomfort with the agitation on board;

– senior managers did not feel troubled by the concerns and heated discussions among drillers;

– excessive amount of mud barrels were brought up during well purges;

– pressure inconsistencies in the drill pipe and in the kill line (pipe through which a vacuum test is performed to ensure that the bottom of the well is sealed);

– the less experienced were misled by a misrepresentation of the change in pressure.

A11.6. The elements noted in [MAR 10b]

– The leak was at its peak from April 20 to June 5, and then continued in a mitigated manner for a few weeks. It was only on August 5 that BP announced the plugging of the well;

– accumulated emissions at sea estimated at nearly 5 million barrels;

– the depth made successive interventions to stop the flow extremely difficult;

– one of the big surprises caused by the accident was the total absence of an adapted operational intervention plan;

– what gives the Deepwater Horizon accident a particular reprieve is the great depth at which it took place and which made the responses extremely complex;

– it is a fact that BP, in order to save time and money, did not implement the best available technologies;

– the explosion prevention system was not equipped with the infrared remote controls used, for example, by Total or Shell in deep offshore;

– drilling was not accompanied by the parallel construction of a relief well, a device made mandatory by Norway, Brazil and Canada for deep offshore operations in sensitive contexts that saves time in the event of an accident and limits the impact, but almost doubles the cost of drilling;

– the human proximity between the oil majors and the federal services responsible for granting drilling permits and monitoring safety was also cited as a reason for the laxity shown by the supervisory authorities in monitoring this high-risk operation;

– the emphasis on economic profitability would explain why the most efficient security devices were not deployed here;

– risk calculation models, based on data from past projects, were translated into accident probabilities and most often came to the conclusion that the risk of explosion (complete loss of control of operations with destruction of infrastructure) was of the order of 10-9; therefore, a shock to a pocket of high-pressure gas not absorbed by the BOP, would almost never occur. However, the BP disaster did occur;

– reasoning based on technological command therefore became of little use in risk management. For risk probability calculations – based on statistics that cannot exist, since each new project in these extreme conditions is new – are a guessing game when it comes to operating in such areas.

Appendix 12

Synthesis Note of the Work of IMdR–AFPCN: "Vulnerability of Networks and Natural Disasters"

Several major climatic events of the last decade (floods, storms, marine submersion) have shown the vulnerability networks that constitute the economic and social fabric of France.

The *Institut pour la Maîtrise des Risques* (IMdR) and the *Association Française de Prévention des Catastrophes Naturelles* (AFPCN) have given a mission to a joint working group to reflect and propose preventive actions to be implemented to reduce this vulnerability and increase the resilience of the various networks concerned.

From four plenary meetings of about 30 people (held between 2009 and 2012) with participation and testimonies of actors representing different networks that have been affected by major climate events, the working group proposed a method of approach that had not been applied so far in the studies of natural hazards.

This method (known as "cindynics") does not focus on the technical shortcomings and the material means to overcome them, but rather on the behavior of the various actors involved (network managers, users, public authorities, etc.).

A comparative analysis of the expected modes of action and those observed in each group of actors, as well as the interactions observed between these groups, makes it possible to identify shortcomings in the behavior of actors and in the implementation of the means made available to them.

The method then proposes the implementation of preventive actions which can certainly include proposals for network development but which, above all, aim to involve the actors concerned in their individual and relational behavior. These actions aim at anticipating and implementing, if need be, provisions that would allow people and goods to be protected, to alleviate the impact of a crisis situation and to move as soon as possible from degraded to normal activity.

Some lessons to be learned from this approach have already been formulated in the case of the study undertaken to be the subject of recommendations to the actors concerned. As an example, we can cite the following:

– for network operators: coordination of the transmission of information by promoting short circuits;

– for the State and communities: a critical analysis of the regulatory and procedural arsenal;

– for departmental fire and rescue services: diversification of information networks and prioritization of missions;

– for economic agents (industry and services): reduction of risk denial and promotion of insurance practices;

– for susceptible establishments: promotion of mobile, removable and easy-to-use protective devices;

– for social network users: restoring a risk culture and raising awareness to make housing and access (connection) to networks less vulnerable.

With a few ideas for the implementation of the proposed actions:

– by involving federating actors: regulatory authorities, professional federations, consulting firms and research actors;

– planning and anticipating crisis management: Municipal Safeguard Plan and Business Continuity Plan requirements and scenario-based work;

– by reducing economic vulnerability: economic analyses integrating all the costs related to the interruption of networks (economic, health and social impacts, environmental impacts, etc.);

– by improving communication between actors: reinforcing the influence of local authorities who often do not control the networks even though they are the conceding authorities.

Appendix 13

The New Cindynics Concepts Training Course

The lessons learned from the work carried out within the framework of the "Cindynics for All Working and Reflection Group" led to the improvement of the cindynics concepts training course, intended to spread the power of these concepts within the science of danger.

This twelve-hour program has been divided into three sessions:

– the importance of cindynics;

– process, method and tools;

– fields of use;

– practice: theory followed by a case study.

The case study is assessed according to seven deliverables:

– system definition and activity situation;

– selecting the concerned and relevant actors;

– characterizing these actors through their visible behavior in reference to expected behavior;

– identifying the dreaded dangers at the global level of the organism;

– identifying the dreaded dangers at the level of groups of actors;

– identifying the dreaded dangers at the level of each actor;

– developing an action plan to reduce the various vulnerabilities of the organism.

At the end of this course, participants are invited to study a case that has occurred in their own organisation in order to help them better understand how the concepts of cindynics can help.

Postface

It is an honor to write the postface of this book, which stems from the "Cindyniques pour tous" working and think tank group of the *Institut pour la Maîtrise des risques* (French Institute for Risk Management), managed by Guy Planchette.

May all authors, proofreaders and active members of the working group be warmly thanked here for this book and what they communicate of their thoughts, professional practices and contributions to the development of the science of danger.

I salute the tireless commitment of the founder and honorary president of the French Institute for Risk Management in understanding and disseminating the work of Georges-Yves Kervern (1935–2008) and those who accompanied him in the still-young history of cindynics.

Guy Planchette and his co-authors give us here an exceptional synthesis informed by both major scientific references and the practices of those whose lives are affected by its approaches and tools. With generosity, they help us to fight deficits and dissonances which have the potential to destroy so many efforts to improve our conditions of survival, life and work, and constitute huge human and material waste.

Pioneers and students who had the good fortune to know Georges-Yves Kervern were able to admire the links he forged between high scientific culture, major industrial responsibilities and the support of various professional organizations.

The "Cindyniques pour tous" working group continues to support this dynamic and helps us to better understand how our organizations, cultures, management models and practices can insidiously create dangerous situations until the risks explode.

For my part, I can attest to the exceptional relevance of cindynic hyperspace applied to the management of hospitals and healthcare establishments, small- and medium-sized enterprises and large organizations in the social economy.

The Covid-19 pandemic has brought to light how inattention to the divergence between goals, values, data, models, laws and standards can weaken all of humanity exposed to a virus.

So, in a way not dissimilar to human biology, the content of this book shows to what extent the cindynics approach primarily auscultates the fragilities of collective activity situations: from cognitive biases to cultural, organizational and managerial deficits.

This way of being and doing develops our intelligence and ability to act and encourages us to positively fight against multiple forms of conflict.

The diagnoses lead to preventive and therapeutic actions. The program for the next 10 years is therefore clear: to move from the sometimes painful identification of threats and weaknesses to that of opportunities and strengths. This is exciting, useful, vital.

What an antidote to the declinist *doxa* contemptuous of the efforts of the people and of the real elites committed to building together, to moving from "power-over" to "power-with", away from the excesses of governance by texts and numbers that destroy professions, their meaning, their learning and their continuous improvement practices!

Following Georges-Yves Kervern, Patrick Rubise, Mioara Mugur-Schächter, Philippe Boulenger, Laurence Baillif and Guy Planchette, we are increasingly learning to better abandon the illusions and false securities of an accumulation of standards and injunctions which end up masking the overall views, even extinguishing the desire for them, under the influence of "infobesity", commentary and polemicism.

What are the real dangers? How to detect them early and what to do, together, to avoid falling too often into the traps set by fallacious representations of reality? This synthesis invites us, in the diverse circles and organizations to which we belong, to appropriate the cindynics which will help us to answer these questions. Nourished by the fields of industry and insurance, they were designed for these purposes.

I hope that Guy Planchette will allow me to underline how motivating his experience, his perception of the essential, his desire to continue the work of

Georges-Yves Kervern and his sense of the future are and how they carry multiple applications, in many fields.

From the conditions and choices of their conception to their implementation, cindynics allow oxygen to circulate. Their practice will nourish our faculties for spotting gaps, deficits, dissonances, disjunctions, degenerations, blockages which produce very concrete vulnerabilities; so we can reduce them, at all scales.

Thus, cindynics, a movement of the mind, an incentive to real exchanges and preventative action, participate in individual, cultural and social vitality.

Georges-Yves Kervern, Guy Planchette and members of "Cindynics for all" from the Institute for Risk Management, national and international organizations which everywhere test the relevance of the science of danger and work for its development, let's learn to breathe that air together!

Jean-Marie FESSLER

Former hospital director and health facilities of the *Mutuelle générale de l'Education Nationale*, France, consulting professor for Stanford University, doctor of Medical Ethics and Health Economics

July 2021

Glossary

Definitions and explanations marked with an asterisk* are taken from the literature developed by Kervern *et al.*

Please note that all elements of the glossary presented below are of a conventional nature only.

Activity situation: situation in which there is a willingness to act.

Actor: an individual, with their own characteristics, who when placed within an organization, becomes an actor through the formal and informal relationships they establish in their work activity.

Aggravating factor: risk factor aggravating the damage generated by an accident (e.g. speeding).

Aspect: in the sense of the relativized conceptualization method, from the point of view of a qualification (e.g. color, length, weight, etc.).

Axiology: the science of moral values; according to the Encyclopedia Universalis, it is the study or theory of what is worthy of esteem, of what is valuable, of what can be the object of a value judgment; how the rules will be applied.

Baseline situation: desired situation within the reference organization and defined by the client of a study.

Blocking: position of refusal, rejection, denial.

Cindynics*: translation from the Greek *kindunos* meaning danger: the science of danger comprising a set of sciences and techniques studying, before those of risks, all forms of danger (or sources of risks), whether they are perceptible or not, in order to better circumscribe them; according to the Larousse, cindynics represents all of the sciences and techniques that study risks (natural, technological, etc.) and their prevention.

Cindynogenesis*: refers to the production of danger.

Cindynogenic*: action that generates danger.

Cindynolysis*: action of danger reduction.

Cindynometry*: statistics of the measurement of dangers and risks.

Complexity: two dimensions characterize complexity, that of "globality" and that of "interaction". For the first dimension, complexity is then the fact of a reality that brings together, in solidarity, several different elements. For the second, it is a reciprocal action or influence between two or more elements of a system or different systems that creates new properties through the effect of emergence.

Dangerous phenomenon: corresponds to a release of energy or substance representing dangers and producing effects likely to inflict damage in regards to certain issues.

Dangerous situation: any situation in which a vulnerable target is exposed to one or more hazards (or sources of risk).

Data: elementary description of a reality. Here, data express facts and statistics.

Deficit: the difference between an actual quantity and a quantity planned or required to meet a certain demand. *A situation of imbalance if the quantity detected is less than the quantity considered optimal.*

Degeneration: lack of order, lack of *hierarchy* leading to a *blur* in the priority of the choices to be made.

Deontology: a set of rules and duties governing a profession.

Dependability: the ability of an entity to function when and as required.

Disjunction: abnormal disunion or separation between two elements usually joined together. Manifest *inconsistency* between two choices located in different spaces of hyperspace.

Dissonance: disagreement between two or more people due to differences in character or point of view. A state of tension in which an individual is confronted with a situation contrary to their convictions or habits of action.

Emergence: property which presents, at the level of a system, a character of novelty compared to those possessed by each of its components considered in isolation.

Entropic propensity: inherent tendency of the change towards energy degradation, towards increased disorder.

Epistemic: expresses everything that relates to the body of knowledge specific to a social group and a period of time.

Ethology: the scientific study of behavior.

Exposure to a hazard: two cases are to be considered:

– the active case, facing a hazard in the hope of gaining an advantage while being aware (or not) of its existence;

– the passive case, where we can suffer the damage caused by a hazard without being able to free ourselves from it (case of public transport passengers).

Feedback: an approach aimed at analyzing any discrepancy (anomaly, incident, accident, etc.), to determine the causes and to define a set of lessons and actions to prevent their repetition.

Gap: lack of knowledge or omission of one or more elements.

Group of actors: a group of individuals composing a situation of activities and having a set of related data, models, rules, values and purposes.

Hazard: defined as any source with intrinsic capabilities that can cause damage, harm or adverse effects to a vulnerable issue (person, material or immaterial property, organization, etc.). This source can be an object, a substance, a situation, a phenomenon, an anomaly, an ambiguity, etc. The ISO 31000:2018 standard uses the term risk source to qualify the hazard.

Hyperspace of danger (HED)*: qualification grid of dangerous situations with five qualifiers (purposes, values, rules, models and data). This grid is used as a tool to obtain a relativized description of impalpable or unobservable hazards.

Issues: people, goods, equipment, environment, likely to be affected by a natural and/or anthropogenic phenomenon and to suffer prejudice or damage.

Models: an artificial system whose properties are similar to observed or inferred properties of a system under study, and whose behavior is intended to either reveal original behaviors that can be the subject of new investigations, or to test to the extent to which the properties attributed to the original can account for its manifested behavior [THI 75]. A model is a form of representing the original.

Networks of actors: networks include **all of the interconnected actors** identified during the description of a situation of activities. Actors can be from companies, administrations, public services, political or institutional spheres, etc., without forgetting the population.

Objective: a result towards which someone's action is directed; an objective is a quantitatively or qualitatively measurable step towards the achievement of an end goal.

Organization: an entity (such as companies, services, administrations, associations, unions, etc.) organized around goal-oriented collective actions; by analogy in biology, an organism refers to a complex and organized living system consisting of a set of goal-oriented functions:

– according to Morin's [MOR 77] definition, it "binds together, in an interactive way, elements, individuals, who become the components of a whole".

Propensity: an inherent tendency to change that is manifested in all entities of the dynamic universe in which we live.

Purpose: the ultimate goal of an organization.

Qualifiers: attributes, aspects or signs allowing for the characterization of something by the attribution of a quality. These are different points of view from which hazards can be described.

Reference organization: organization considered "ideal" or desired by a sponsor.

Resilience: generally refers to the ability of an organization, group or structure to adapt to a changing environment (ISO Guide 73:2009). It is used in several contexts (physical, technical, psychological, ecological, economic, etc.).

Risk: this term has a double meaning, one negative meaning referring to a hazard and the other to success after taking a risk; in the case of exposure to a hazard, it usually refers to the probability of suffering harm or adverse health effects; ISO 31000 defines it as "the effect of uncertainty on objectives", referring to the double meaning.

Risk factor: a condition, behavior, catalyst or deficit, which increases the likelihood that an undesired event will occur (failure to respect distances between vehicles).

Robustness: a characteristic of what presents a constitution resistant to the causes of aggression or alteration.

Rule: a principle of living in society. Here, the notion of rule applies to laws, regulations, standards, procedures, etc.

Safety: resistance to one or more unauthorized intentional acts intended to cause damage (to persons, property or the environment) or to interfere with the operation of an entity.

Security: the absence of unacceptable risk to people, property or the environment from threats.

Situation: in the etymological sense of the term: the location of a building and according to the Larousse, the set of events, circumstances and concrete relationships in the middle of which a person or a group finds themselves.

Stakeholders: according to the ISO Guide 73: 2009: a person or organization that can affect, be affected by, or perceive themselves to be affected by a decision or activity.

System: According to de Rosnay's definition, a system is characterized by:

– a purpose, corresponding to an economic, social or human utility;

– an organized structure, composed of a large number of elements or sub-systems (technical, human, natural entities);

– interactions resulting in an infinite number of exchanges (energy, information, etc.).

Systemic cindynogenic deficits*: a set of general causes explaining how the cultural, organizational and managerial contexts created within an organization can shape the behavior of the actors involved.

Transformation operators: set of actions transforming a baseline situation into a new one. These actions may be unintentional, acting as incidental, accidental or even catastrophic transformers. They may be intentional, presenting either a threat from outside the system, or an internal desire to lead change.

Uncertainty: in physics, the margin of imprecision on the value of the measurement of a physical quantity.

Values: physical, intellectual and moral qualities of a person that makes them worthy of esteem. Values correspond to what people attribute importance to: such as sharing, equity, justice, honor, etc.

Vulnerability: the character of what is precarious, what can be attacked, injured, damaged. It is therefore a sensitivity to a type of positive or harmful effect that applies to people, objects or systems (companies, ecosystems, etc.).

References

[ALE 09] ALEXANDRE-BAILLY F., BOURGEOIS D., GRUERE J.P. *et al.*, *Comportements humains et management*, 3rd edition, Pearson, Paris, 2009.

[BAE 10] BAELI R., *Les Archéocindyniques – Analyse historique d'événements cindyniques*, Cindynica, Marseille, 2010.

[BAQ 07] BAQUIAST J.-P., *Pour un principe matérialiste fort*, Éditions Jean-Paul Bayol, Alès, 2007.

[BAR 77] BARDIN L., *L'analyse de contenu*, PUF, Paris, 1977.

[BAS 02] BASDEVANT J.-L., DALIBARD J., Mécanique quantique, Cours à l'Ecole Polytechnique, available at https://www.psychaanalyse.com/pdf/MECANIQUE%20QUANTIQUE%20-%20COURS%20DE%20L%20ECOLE%20POLYTECHNIQUE%202002%20(%20516%20Pages%20-%207,4%20Mo).pdf, February 2002.

[BEA 07] BEAULIEU S., COUILLARD M., "Avis scientifiique sur le port du masque dans la communauté en situation de pandémie d'influenza", Institut National de Santé Publque du Québec, 2007.

[BEL 95] BELIS M., *Causalité, propension, probabilité*, Intellectica, France, 1995.

[BER 16] BERG P., *Deepwater Horizon* [Film], Lionsgate, 2016.

[BOU 20] BOURGUIBA L., "Turbulent gas clouds and respiratory pathogen emissions: Potential implications for reducing transmission of COVID-19", *JAMA*, 2020.

[CAM 21] CAMBRIDGE DICTIONARY, Disjunction, available at: https://dictionary.cambridge.org/dictionary/english/disjunction, 2021.

[CHA 05] CHARBONNIER J., *Bhopal; le pire catastrophe industrielle de tous les temps*, Préventique, Bordeaux, 2005.

[COH 13] COHET P., Disparités de perception et divergences prospectives : prévention et résolution de conflits, maîtrise des risques et développement, IFREI, available at: https://www.ifri.org/en/publications, 2013.

[DAN 10] DANIELOU F., SIMARD M., BOISSIERES I., *Facteurs humains et organisationnels de la sécurité industrielle : un état de l'art*, Les Cahiers de la sécurité industrielle, FONCSI, Toulouse, 2010.

[DEE 11] DEEP HORIZON STUDY GROUP, Final report on the investigation of the Macondo well blowout, available at: http://large.stanford.edu/courses/2011/ph240/mina1/docs/DHSGFinalReport-March2011-tag.pdf, 2011.

[DEL 20] DELHOMMAIS P.-A., Les ehpad contre le CAC 40, *Le Point*, available at: https://www.lepoint.fr/editos-du-point/pierre-antoine-delhommais/pierre-antoine-delhommais-les-ehpad-contre-le-cac-40--19-03-2020-2367854_493.php, March 19, 2020.

[DON 03] DONNADIEU G., DURAND D., NEEL D. *et al.*, L'approche systémique : de quoi s'agit-il ?, available at: www.afscet.asso.fr/SystemicApproach, 2003.

[DUB 16] DUBAU L., MAGNE J., PIGNAULT J., "Effet doppler et cindyniques. Comment concilier signal et mouvement pour maîtriser les risques", *20ème Congrès de Maîtrise des risques et de sûreté de fonctionnement*, Saint-Malo, 2016.

[FES 09] FESSLER J-M., *Cindyniques et santé – contribution des sciences du danger à la santé*, Économica, Paris, 2009.

[FOU 97] FOURNIER A., GUITTON C., KERVERN G-Y. *et al.*, *Le risque psychologique majeur*, Éditions ESKA, Paris, 1997.

[FRA 04] FRANTZEN C., *Tango on Asymptote. Présentation à la 13ème conférence annuelle SRA Europe*, Paris, November 15–17, 2004.

[HAN 17] HANSEN J.-C., PLANCHETTE G., "Intérêt des cindyniques dans le pilotage stratégique des crises et catastrophes", *Manuel de médecine de catastrophe*, Lavoisier, Paris, 2017.

[HUM 10] HUMPHREY A.-S., *Management Consulting, Transformation of Culture, SWOT Analysis*, Harvard University, SRI International, Investors in People, Cambridge, MA, 2010.

[JUL 20] JULIEN H., HELFER B., HANSEN J.-C., "Usage et mésusage des masques alternatifs – place des masques faits maison", *SFMC*, Paris, 2020.

[KER 91] KERVERN G.-Y., RUBISE P., *L'archipel du danger – introduction aux cindyniques*, Économica, Paris, 1991.

[KER 95] KERVERN G-Y., *Éléments fondamentaux des cindyniques*, Économica, Paris, 1995.

[KER 07] KERVERN G.-Y., BOULENGER P., *Cindyniques – concepts et modes d'emploi*, Économica, Paris, 2007.

[KER 11] KERMISH C., *Le concept du risque. De l'épistémologie à l'éthique*, Lavoisier, Paris, 2011.

[LAB 87] LABORIT H., *Dieu ne joue pas aux dés*, Edition Grasset, Paris, 1987.

[LAG 83] LAGADEC, *Le Risque technologique majeur et les situations de crise*, Groupe de Prospective, Secrétariat d'État à l'Environnement, Ministère de l'Urbanisme et du Logement, 1983.

[LAN 20] LANDIER A., THESMAR D., Le Covid, la science et la douloureux apprentissage de l'incertitude, *Les Échos*, available at: https://www.lesechos.fr/idees-debats/cercle/le-covid-la-science-et-le-douloureux-apprentissage-de-lincertitude-1210596, 2020.

[LAP 01] LAPIERRE D., MORO J., *Il est minuit cinq à Bhopal*, Laffont, Paris, 2001.

[LEB 00] LE BRAS H., *Naissance de la mortalité. L'origine politique de la statistique et de la démographie*, Hautes Etudes, Gallimard/Le Seuil, Paris, 2000.

[LEC 02] LE COZE J.-C., LIM S., DECHY N. *Intégration des aspects organisationnels dans le retour d'expérience*, Inéris, Paris, 2002.

[LEC 16] LE COZE J.-C., *Trente ans d'accidents : le nouveau visage des risques sociotechniques*, Octarès, Toulouse, 2016.

[LIG 06] LIGERON J.-C., *Le Cercle des fiabilistes disparus*, Éditions Préventique, Bordeaux, 2006.

[LLO 10] LLORY M., MONTMAYEUL R., *L'accident et l'organisation*, Groupe Préventique, Bordeaux, 2010.

[LLO 18] LLORY M., MONTMAYEUL R., ASSOCIATION CHAOS, *Comprendre l'accident – La pratique de l'analyse organisationnelle de la sécurité*, L'Harmattan, Paris, 2018.

[MAG 10a] MAGRIN G., VAN DESSEL B., *BP "Deepwater Horizon" du Golfe du Mexique à l'Afrique : un tournant pour l'industrie pétrolière ?*, EchoGéo, Sur le Vif, France, 2010.

[MAG 10b] MAGRIN, G., VAN DESSEL B., BP 'Deepwater Horizon' du Golfe du Mexique à l'Afrique : un tournant pour l'industrie pétrolière ?, *EchoGéo*, available at: https://journals.openedition.org/echogeo/12099?lang=en#toc, 2010.

[MAG 17] MAGNE J., PIGNAULT J., DUBAU L., *La prise de décision agile : anticiper les risques grâce aux signaux précurseurs*, Dunod, Paris, 2017.

[MAY 20] MAYER N., Coronavirus : tout ce que l'on sait sur la dispersion des gouttelettes, *Futura Santé*, available at: https://www.futura-sciences.com/sante/actualites/coronavirus-coronavirus-tout-ce-on-sait-dispersion-gouttelettes-80375/, 2020.

[MON 13] MONTMAYEUL R., *L'accident de l'usine chimique de Flixborough*, Groupe Préventique, Bordeaux, 2013.

[MOR 77] MORIN E., *Méthode – la nature de la nature*, Le Seuil, Paris, 1977.

[MUC 91] MUCCHIELLI A., *Les méthodes qualitatives*, PUF, Paris, 1991.

[MUG 77] MUGUR-SCHÄCHTER M., "Les leçons de la mécanique quantique : vers une épistémologie formalisée", *Revue le Débat*, no. 94, 1977.

[MUG 02] MUGUR-SCHÄCHTER A., "Quantum mechanics versus a method of relativized conceptualization", MUGUR-SCHÄCHTER M., VAN DER MERWE A. (eds), *Quantum Mechanics, Mathematics, Cognition and Action: Proposals for a Formalized Epistemology*, Springer, Heidelberg, 2002.

[MUG 06] MUGUR SCHÄCHTER M., *Sur le tissage des connaissances*, Hermes-Lavoisier, Paris, 2006.

[NIC 12] NICOLET J-L., *Le risque technologique majeur à l'épreuve du droit*, vol. 2, L'Harmattan, Paris, 2012.

[PAI 94] PAILLÉ P., "L'analyse par théorisation ancrée", *Cahiers de recherche sociologique*, no. 23, pp. 147–181, Erudit, Montreal, 1994.

[PAI 96] PAILLÉ P., "De l'analyse qualitative en général et de l'analyse thématique en particulier", *Recherches Qualitatives*, no. 15, pp. 179–194, Montreal, 1996.

[PAU 95] PAUCHANT T.C., MITROFF I.I., *La gestion des crises et des paradoxes*, Éditions Québec-Amérique, Montreal, 1995.

[PÉR 00] PÉRILHON P., "MOSAR : présentation de la méthode. Sécurité et gestion des risques", *Techniques de l'Ingénieur*, Article SE4060, pp. 1–16, 2000.

[PNR 17] PNRS, Retour d'expérience sur les accidents industriels, *Proceedings of the 12th seminar May–June 2017*, available at: https://www.impel.eu/wp-content/uploads/2015/08/Brochure_IMPEL2017_Fr.pdf, 2017.

[POP 91] POPPER K., *Un univers de propension*, Éditions de l'Éclat, Paris, 1991.

[RAS 84] RASMUSSEN J., PEDERSEN O.M., Human factors in probabilistic risk analysis and in risk management, Report, International Nuclear Information System, Vienna, 1984.

[RAS 86] RASMUSSEN J., *Information Processing and Human-Machine Interaction*, North-Holland, New York, 1986.

[RAS 97] RASMUSSEN J., *Risk Management in a Dynamic Society: A Modelling Problem*, Elsevier Safety Science, Amsterdam, 1997.

[REA 91] REASON J., *Human Error*, Cambridge University Press, Cambridge, 1991.

[REA 13] REASON J., *L'erreur humaine*, Presses des Mines, Paris, 2013.

[SHI 12] SHIMAZU N., The Fukushima report hides behind the cultural curtain, *The Guardian*, available at: https://www.theguardian.com/commentisfree/2012/jul/06/fukushima-report-disaster-japan, 2012.

[STE 04] STETZENBACH L.D., BUTTNER M.P., CRUZ P., "Détection et dénombrement des biocontaminants aéroportés", *Opinion actuelle en biotechnologie*, vol. 15, no. 3, pp. 170–174, 2004.

[THI 75] THINES G., LEMPEREUR A., *Dictionnaire Général des sciences humaines*, Éditions Universitaires, Paris, 1975

[VAU 96] VAUGHAN D., *The Challenger Launch Decision: Risky Technology, Culture and Deviance at NASA*, University of Chicago, Chicago, IL, 1996.

[VIL 88] VILLEMEUR A., *Sûreté de fonctionnement des systèmes industriels. Collection de la Direction des études et recherches d'EDF*, Eyrolles, Paris, 1988.

Other documents in the IMdR library

IMdR λμ 21 conference (2018)

[LAN 18] LANNOY A., PLANCHETTE G., *Conférence introductive : de la sûreté de fonctionnement à la maîtrise des risques : un regard sur le passé pour mieux se projeter dans l'avenir*, IMdR, Gentilly, 2018.

Proceedings (IEC)

CINDYNICS 92 – "*Cindyniques*", Cannes, 1992.

CINDYNICS 94 – "*Risque et société*", Paris-La Sorbonne, November 1994.

CINDYNICS 97 – "*La violence est-elle un accident ?*", Paris-La Sorbonne, 1997.

*Notebooks from the "***Épistémologie des cindyniques***" group*

N° 1 June 1995: La fonction sociale de la confiance : action collective et délégation de responsabilité face au risque.

N° 2 January 1996: Du danger au risque : économie des transactions dans la régulation sociale.

N° 3 December 1996: Risque, santé, environnement : la place de l'outil économique dans la transaction sociale.

N°4 January 1998: Une pragmatique de la précaution : du rationnel au raisonnable.

N° 5 November 1999: L'irréversibilité dans la gestion du risque : la réversibilité comme instrument de précaution.

Index

R, S

T, V

Other titles from

ISTE

in

Mechanical Engineering and Solid Mechanics

2021

CHALLAMEL Noël, KAPLUNOV Julius, TAKEWAKI Izuru
Modern Trends in Structural and Solid Mechanics 1: Static and Stability
Modern Trends in Structural and Solid Mechanics 2: Vibrations
Modern Trends in Structural and Solid Mechanics 3: Non-deterministic Mechanics

DAHOO Pierre Richard, POUGNET Philippe, EL HAMI Abdelkhalak
Applications and Metrology at Nanometer Scale 1: Smart Materials, Electromagnetic Waves and Uncertainties (Reliability of Multiphysical Systems Set – Volume 9)
Applications and Metrology at Nanometer Scale 2: Measurement Systems, Quantum Engineering and RBDO Method (Reliability of Multiphysical Systems Set – Volume 10)

LEDOUX Michel, EL HAMI ABDELKHALAK
Heat Transfer 1: Conduction (Mathematical and Mechanical Engineering Set – Volume 9)
Heat Transfer 2: Radiative transfer (Mathematical and Mechanical Engineering Set – Volume 10)

2020

SALENÇON Jean
Elastoplastic Modeling

2019

BAYLE Franck
Reliability of Maintained Systems Subjected to Wear Failure Mechanisms: Theory and Applications
(Reliability of Multiphysical Systems Set – Volume 8)

BEN KAHLA Rabeb, BARKAOUI Abdelwahed, MERZOUKI Tarek
Finite Element Method and Medical Imaging Techniques in Bone Biomechanics (Mathematical and Mechanical Engineering Set – Volume 8)

IONESCU Ioan R., QUEYREAU Sylvain, PICU Catalin R., SALMAN Oguz Umut
Mechanics and Physics of Solids at Micro- and Nano-Scales

LE VAN Anh, BOUZIDI Rabah
Lagrangian Mechanics: An Advanced Analytical Approach

MICHELITSCH Thomas, PÉREZ RIASCOS Alejandro, COLLET Bernard, NOWAKOWSKI Andrzej, NICOLLEAU Franck
Fractional Dynamics on Networks and Lattices

SALENÇON Jean
Viscoelastic Modeling for Structural Analysis

VÉNIZÉLOS Georges, EL HAMI Abdelkhalak
Movement Equations 5: Dynamics of a Set of Solids
(Non-deformable Solid Mechanics Set – Volume 5)

2018

BOREL Michel, VÉNIZÉLOS Georges
Movement Equations 4: Equilibriums and Small Movements
(Non-deformable Solid Mechanics Set – Volume 4)

FROSSARD Etienne
Granular Geomaterials Dissipative Mechanics: Theory and Applications in Civil Engineering

RADI Bouchaib, EL HAMI Abdelkhalak
Advanced Numerical Methods with Matlab® 1: Function Approximation and System Resolution
(Mathematical and Mechanical Engineering SET – Volume 6)
Advanced Numerical Methods with Matlab® 2: Resolution of Nonlinear, Differential and Partial Differential Equations
(Mathematical and Mechanical Engineering SET – Volume 7)

SALENÇON Jean
Virtual Work Approach to Mechanical Modeling

2017

BOREL Michel, VÉNIZÉLOS Georges
Movement Equations 2: Mathematical and Methodological Supplements
(Non-deformable Solid Mechanics Set – Volume 2)
Movement Equations 3: Dynamics and Fundamental Principle
(Non-deformable Solid Mechanics Set – Volume 3)

BOUVET Christophe
Mechanics of Aeronautical Solids, Materials and Structures
Mechanics of Aeronautical Composite Materials

BRANCHERIE Delphine, FEISSEL Pierre, BOUVIER Salima, IBRAHIMBEGOVIC Adnan
From Microstructure Investigations to Multiscale Modeling: Bridging the Gap

CHEBEL-MORELLO Brigitte, NICOD Jean-Marc, VARNIER Christophe
From Prognostics and Health Systems Management to Predictive Maintenance 2: Knowledge, Traceability and Decision
(Reliability of Multiphysical Systems Set – Volume 7)

EL HAMI Abdelkhalak, RADI Bouchaib
Dynamics of Large Structures and Inverse Problems

(Mathematical and Mechanical Engineering Set – Volume 5)
Fluid-Structure Interactions and Uncertainties: Ansys and Fluent Tools
(Reliability of Multiphysical Systems Set – Volume 6)

KHARMANDA Ghias, EL HAMI Abdelkhalak
Biomechanics: Optimization, Uncertainties and Reliability
(Reliability of Multiphysical Systems Set – Volume 5)

LEDOUX Michel, EL HAMI Abdelkhalak
Compressible Flow Propulsion and Digital Approaches in Fluid Mechanics
(Mathematical and Mechanical Engineering Set – Volume 4)
Fluid Mechanics: Analytical Methods
(Mathematical and Mechanical Engineering Set – Volume 3)

MORI Yvon
Mechanical Vibrations: Applications to Equipment

2016

BOREL Michel, VÉNIZÉLOS Georges
Movement Equations 1: Location, Kinematics and Kinetics
(Non-deformable Solid Mechanics Set – Volume 1)

BOYARD Nicolas
Heat Transfer in Polymer Composite Materials

CARDON Alain, ITMI Mhamed
New Autonomous Systems
(Reliability of Multiphysical Systems Set – Volume 1)

DAHOO Pierre Richard, POUGNET Philippe, EL HAMI Abdelkhalak
Nanometer-scale Defect Detection Using Polarized Light
(Reliability of Multiphysical Systems Set – Volume 2)

DE SAXCÉ Géry, VALLÉE Claude
Galilean Mechanics and Thermodynamics of Continua

DORMIEUX Luc, KONDO Djimédo
Micromechanics of Fracture and Damage
(Micromechanics Set – Volume 1)

EL HAMI Abdelkhalak, RADI Bouchaib
Stochastic Dynamics of Structures
(Mathematical and Mechanical Engineering Set – Volume 2)

GOURIVEAU Rafael, MEDJAHER Kamal, ZERHOUNI Noureddine
From Prognostics and Health Systems Management to Predictive Maintenance 1: Monitoring and Prognostics
(Reliability of Multiphysical Systems Set – Volume 4)

KHARMANDA Ghias, EL HAMI Abdelkhalak
Reliability in Biomechanics
(Reliability of Multiphysical Systems Set –Volume 3)

MOLIMARD Jérôme
Experimental Mechanics of Solids and Structures

RADI Bouchaib, EL HAMI Abdelkhalak
Material Forming Processes: Simulation, Drawing, Hydroforming and Additive Manufacturing
(Mathematical and Mechanical Engineering Set – Volume 1)

2015

KARLIČIĆ Danilo, MURMU Tony, ADHIKARI Sondipon, MCCARTHY Michael
Non-local Structural Mechanics

SAB Karam, LEBÉE Arthur
Homogenization of Heterogeneous Thin and Thick Plates

2014

ATANACKOVIC M. Teodor, PILIPOVIC Stevan, STANKOVIC Bogoljub, ZORICA Dusan
Fractional Calculus with Applications in Mechanics: Vibrations and Diffusion Processes
Fractional Calculus with Applications in Mechanics: Wave Propagation, Impact and Variational Principles

CIBLAC Thierry, MOREL Jean-Claude
Sustainable Masonry: Stability and Behavior of Structures

ILANKO Sinniah, MONTERRUBIO Luis E., MOCHIDA Yusuke
The Rayleigh–Ritz Method for Structural Analysis

LALANNE Christian
Mechanical Vibration and Shock Analysis – 5-volume series – 3rd edition
Sinusoidal Vibration – Volume 1
Mechanical Shock – Volume 2
Random Vibration – Volume 3
Fatigue Damage – Volume 4
Specification Development – Volume 5

LEMAIRE Maurice
Uncertainty and Mechanics

2013

ADHIKARI Sondipon
Structural Dynamic Analysis with Generalized Damping Models: Analysis

ADHIKARI Sondipon
Structural Dynamic Analysis with Generalized Damping Models: Identification

BAILLY Patrice
Materials and Structures under Shock and Impact

BASTIEN Jérôme, BERNARDIN Frédéric, LAMARQUE Claude-Henri
Non-smooth Deterministic or Stochastic Discrete Dynamical Systems: Applications to Models with Friction or Impact

EL HAMI Abdelkhalak, RADI Bouchaib
Uncertainty and Optimization in Structural Mechanics

KIRILLOV Oleg N., PELINOVSKY Dmitry E.
Nonlinear Physical Systems: Spectral Analysis, Stability and Bifurcations

LUONGO Angelo, ZULLI Daniele
Mathematical Models of Beams and Cables

SALENÇON Jean
Yield Design

2012

DAVIM J. Paulo
Mechanical Engineering Education

DUPEUX Michel, BRACCINI Muriel
Mechanics of Solid Interfaces

ELISHAKOFF Isaac *et al.*
Carbon Nanotubes and Nanosensors: Vibration, Buckling and Ballistic Impact

GRÉDIAC Michel, HILD François
Full-Field Measurements and Identification in Solid Mechanics

GROUS Ammar
Fracture Mechanics – 3-volume series
Analysis of Reliability and Quality Control – Volume 1
Applied Reliability – Volume 2
Applied Quality Control – Volume 3

RECHO Naman
Fracture Mechanics and Crack Growth

2011

KRYSINSKI Tomasz, MALBURET François
Mechanical Instability

SOUSTELLE Michel
An Introduction to Chemical Kinetics

2010

BREITKOPF Piotr, FILOMENO COELHO Rajan
Multidisciplinary Design Optimization in Computational Mechanics

DAVIM J. Paulo
Biotribolgy

PAULTRE Patrick
Dynamics of Structures

SOUSTELLE Michel
Handbook of Heterogenous Kinetics

2009

BERLIOZ Alain, TROMPETTE Philippe
Solid Mechanics using the Finite Element Method

LEMAIRE Maurice
Structural Reliability

2007

GIRARD Alain, ROY Nicolas
Structural Dynamics in Industry

GUINEBRETIÈRE René
X-ray Diffraction by Polycrystalline Materials

KRYSINSKI Tomasz, MALBURET François
Mechanical Vibrations

KUNDU Tribikram
Advanced Ultrasonic Methods for Material and Structure Inspection

SIH George C. *et al.*
Particle and Continuum Aspects of Mesomechanics